Why Smart Students Feel Stuck

A practical roadmap that shows high-ability students how to break hidden mental traps, rebuild momentum, and turn intelligence into consistent progress and real-world results.

By Laura Lee

Lauxon Publishing, LLC

Published by Lauxon Publishing, LLC

For permission requests, please contact:
Email: info@Lauxonpublishing.com
Website: www.Lauxonpublishing.com

Author's Note

I wrote this book because I see you—the student, the thinker, the achiever who feels like intelligence is both a gift and a burden. I know what it's like to feel stuck in your own mind, to measure yourself against invisible standards, and to hesitate for fear of "not living up to potential."

This book is my invitation to step off that treadmill and explore a different approach: one where progress matters more than perfection, where curiosity fuels action, and where self-trust becomes the engine of growth. Every chapter reflects not only research and experience but also a belief that growth is not linear, intelligence is not identity, and setbacks are opportunities waiting to be reframed.

Consider this book a toolbox. Some tools will feel familiar, others may challenge you—but all are designed to help you move forward, take ownership of your learning, and reclaim momentum in life and study. My hope is that by the last page, you'll not only understand why you feel stuck, but also know exactly how to get unstuck and keep moving forward with confidence.

Table of Contents

PROLOGUE

Somewhere along the path of intelligence and early success, many of us hit an invisible wall. It isn't a lack of talent, ambition, or opportunity—it's the quiet struggle of being smart in a world that often measures worth by speed, perfection, and comparison. For students, the brightest among us, success can feel like a trap: every expectation becomes a mirror, reflecting pressure, doubt, and fear of failure.

This book was born from that tension—the desire to understand why intelligence sometimes feels like a burden, and to discover the keys to turning potential into momentum. It's not about becoming smarter; it's about learning how to think, act, and move forward when the usual markers of success no longer work. It's about transforming overthinking into action, perfectionism into progress, and fear into strategic courage.

Inside these pages, you will find tools, strategies, and perspectives designed to help smart students—and anyone who has ever felt "stuck"—rebuild confidence, take control of learning, and create a life that aligns with curiosity, creativity, and purpose. This is a guide, a companion, and a roadmap for turning intelligence into momentum, and pressure into empowerment.

Part I: Understanding the Invisible Problem

CHAPTER .1

INTELLIGENCE ISN'T THE ADVANTAGE YOU WERE PROMISED

From a young age, being labeled smart often feels like receiving a permanent advantage in life. Teachers praise quick answers, adults predict bright futures, and peers quietly step aside as if intelligence guarantees success. Over time, this label begins to shape expectations not only from others but from within. The assumption forms that intelligence alone will open doors and smooth paths. Effort seems optional when understanding comes quickly. What no one explains is that intelligence can quietly become a fragile foundation if it is never reinforced with skills, habits, and resilience.

The early rewards of being smart can create a distorted relationship with challenge. When tasks are easy, progress feels natural and almost automatic. Difficulty becomes unfamiliar rather than instructional. As a result, intelligence becomes associated with ease instead of persistence. When effort finally becomes necessary, it can feel like something has gone wrong. This moment often marks the beginning of feeling stuck, confused, and disappointed despite clear ability.

Smart students are rarely warned that intelligence is not a strategy. It does not teach how to recover from failure, manage uncertainty, or navigate ambiguity. Intelligence can process information, but it does not automatically build judgment. It can identify patterns, but it does not guarantee follow-through. Without deliberate development, intelligence becomes a tool that is impressive but underutilized. Like any tool, its value depends on how it is applied over time.

Expectations attached to intelligence often create invisible pressure. Being smart becomes something to protect rather than something to use. Risk feels dangerous because mistakes threaten identity. Growth becomes conditional on maintaining an image of competence. Instead of exploring freely, many smart students begin to narrow their choices. They choose paths where they are most likely to succeed quickly and avoid situations where they might struggle publicly.

This pressure quietly shifts motivation from curiosity to performance. Learning becomes about maintaining status rather than expanding ability. Questions are filtered through the fear of appearing uninformed. Effort is hidden to preserve the illusion of ease. Over time, this leads to stagnation disguised as stability. The student may look successful from the outside while feeling increasingly disconnected from growth on the inside.

Intelligence was never meant to be a guarantee. It is a starting point, not a destination. When treated as an identity, it becomes fragile. When treated as a resource, it becomes powerful. The difference lies in whether intelligence is used to avoid struggle or to engage with it. Growth begins when intelligence is allowed to be imperfect, unfinished, and challenged.

Many smart students internalize the belief that needing help means they are failing. They hesitate to ask questions once praise has taught them they should already know. This silence compounds confusion and reinforces isolation. Over time, the gap between perceived ability and actual progress widens. What feels like a personal flaw is often a learned response to unrealistic expectations. The problem is not intelligence itself, but the story built around it.

The world beyond early education rarely rewards raw intelligence alone. It rewards adaptability, communication, persistence, and emotional regulation. These skills develop through friction, not avoidance. When smart students rely solely on cognitive strength, they are often unprepared for environments that demand collaboration and uncertainty. Intelligence without practice becomes brittle under pressure. Strength without flexibility eventually breaks.

Feeling stuck is not a sign that intelligence has failed. It is often a sign that intelligence has been carrying too much weight by itself. Growth requires learning how to struggle productively. It requires redefining success as movement rather than ease. When intelligence is paired with humility and persistence, it regains its usefulness. The stuck feeling becomes a signal rather than a sentence.

There is relief in realizing that intelligence was never supposed to do everything. It does not need to justify worth or predict outcomes. It is allowed to coexist with confusion and effort. When expectations loosen, curiosity returns. Progress resumes when intelligence becomes a partner to action rather than a shield from discomfort. This shift marks the true beginning of growth.

Understanding this invisible problem reframes the entire journey ahead. Being smart did not fail you. The promise attached to it was incomplete. Once intelligence is freed from unrealistic expectations, it can finally be used fully. The path forward begins not by becoming smarter, but by becoming more willing to engage with uncertainty. That willingness is what transforms intelligence into momentum.

The promise of intelligence often comes with an unspoken timeline. Smart students are expected to know early, choose quickly, and succeed fast. When clarity does not arrive on schedule, anxiety replaces confidence. Questions that once felt exciting begin to feel threatening. The gap between expectation and reality grows wider with each passing year. Instead of being curious about that gap, many students judge themselves for it. This judgment quietly drains motivation and self-trust.

Because intelligence was rewarded early, struggle can feel unfamiliar and personal. When progress slows, smart students often assume something is wrong with them rather than with the approach they are using. They may work harder in ineffective ways, doubling down on thinking instead of experimenting. Effort becomes invisible because it is happening internally rather than through action. This creates exhaustion without results. The mind stays busy while life remains unchanged.

Another hidden issue is that intelligence is often measured by speed. Quick comprehension is praised more than deep understanding. Over time, smart students may rush through learning to maintain that identity. Slowing down feels like regression rather than refinement. Depth is sacrificed for efficiency, and reflection is replaced with performance. Eventually, speed stops being enough, and there is no framework for adapting.

The environments that reward intelligence early on are often highly structured. Clear rules, defined problems, and known answers create a sense of safety. Outside those environments, ambiguity dominates. There are no rubrics for life decisions or guaranteed correct answers. Smart students who thrived in structure can feel lost without it. The absence of clear markers makes intelligence feel suddenly unreliable.

This transition exposes a critical misunderstanding about success. Intelligence can solve problems that are well-defined. It struggles when problems are messy, emotional, or ill-structured. Many of life's most important challenges fall into that second category. Relationships, careers, purpose, and meaning cannot be solved like equations. Without practice navigating ambiguity, even very smart people can feel paralyzed.

Feeling stuck is often the first honest feedback smart students receive. It signals that the old rules no longer apply. The strategies that once worked are no longer sufficient. This moment is not a failure but an invitation to evolve. Intelligence must learn to collaborate with patience, experimentation, and resilience. When that collaboration begins, the feeling of being stuck slowly gives way to forward motion.

The belief that intelligence should prevent struggle creates a quiet fear of exposure. Smart students often worry that difficulty will reveal them as less capable than others believe. This fear encourages hiding confusion instead of addressing it. Silence replaces questions, and avoidance replaces engagement. Over time, this behavior reinforces the idea that struggle is shameful. What begins as self-protection slowly becomes self-sabotage.

Another consequence of relying on intelligence is the underdevelopment of learning stamina. When things come easily, there is little incentive to build endurance. Long-term projects, delayed rewards, and repetitive practice feel unusually draining. The mind may be sharp, but the patience muscle is weak. This imbalance becomes visible when progress requires sustained effort. Intelligence alone cannot compensate for the absence of consistency.

Smart students are also more likely to internalize comparison. Praise often positions them as exceptional relative to others. This framing turns peers into benchmarks rather than collaborators. Progress becomes about staying ahead instead of moving forward. Comparison narrows focus and increases anxiety. Growth slows when attention shifts away from learning and toward ranking.

The pressure to appear competent can distort decision-making. Choices are filtered through how they might look rather than what they might teach. Safer options feel more appealing than exploratory ones. This leads to predictable paths that limit discovery. Over time, regret replaces curiosity. The student wonders how they became bored while still being capable.

Relearning how to use intelligence requires reframing its purpose. Intelligence works best when it supports action instead of replacing it. It should help generate options, not eliminate risk. When paired with experimentation, intelligence becomes adaptive. The willingness to look foolish temporarily unlocks long-term growth. This shift transforms intelligence from a fragile label into a flexible tool.

The longer intelligence is treated as something to protect, the more fragile it becomes. Smart students may avoid situations that test them deeply, preferring environments where they can remain competent. This avoidance limits feedback, which is essential for growth. Without feedback, blind spots remain hidden and skills plateau. What feels like maintaining strength is often preventing development. True capability is built by exposure, not preservation.

There is also a subtle grief that accompanies realizing intelligence is not enough. Many students feel betrayed by a system that promised smooth success. This disappointment can turn inward and become self-blame. Instead of questioning the promise, they question their worth. That emotional weight makes it harder to re-engage with challenges. Understanding this grief is a necessary step toward releasing it.

Intelligence was rewarded for answers, not for asking better questions. As a result, smart students may struggle to navigate uncertainty. They look for the correct move rather than the next useful one. This search for certainty delays progress. In complex situations, clarity emerges through action, not contemplation. Intelligence must learn to tolerate incomplete information.

Progress resumes when intelligence is allowed to be unfinished. Smart students often wait until they feel fully prepared before acting. This readiness never arrives because it is built through experience. Action creates the data that thinking alone cannot.

Once this cycle is understood, movement becomes less intimidating. Intelligence shifts from prediction to adaptation.

The real advantage of intelligence is not speed or accuracy, but learning capacity. When that capacity is activated through experience, growth accelerates. Feeling stuck is not a verdict but a transition point. It signals that intelligence is ready to mature. The path forward begins by allowing intelligence to work alongside effort, uncertainty, and time.

The way intelligence is praised often emphasizes outcomes rather than process. Smart students are lauded for getting things right, finishing quickly, or producing exceptional results. Rarely are they praised for persistence, exploration, or resilience. This skewed feedback teaches that the value of intelligence lies in proving it, not using it. As a result, effort becomes secondary and often invisible. Learning to appreciate process is essential to moving beyond feeling stuck.

Another hidden challenge is that intelligence can mask gaps in skill. Early success allows students to bypass areas of weakness. Teachers and peers may assume mastery exists where it does not. These unaddressed gaps compound over time, making later challenges feel insurmountable. The student may wonder why competence feels uneven despite natural ability. Recognizing these gaps is the first step toward meaningful improvement.

Smart students also tend to over-rely on analytical thinking. When faced with messy, emotional, or ambiguous problems, analysis alone is insufficient. Relationships, career navigation, and personal growth cannot always be solved with logic. Over-analysis delays action and increases self-doubt. Learning to balance thinking with doing becomes crucial. Intelligence must become a guide, not a gatekeeper.

Pressure from others adds a silent burden. Teachers, parents, and peers project expectations onto a student labeled "smart." These projections can feel like obligations rather than guidance. Choices are filtered through anticipated judgment rather than curiosity or interest. This external pressure magnifies internal stress and heightens the fear of failure. Over time, it shapes behavior more than personal desire or intrinsic motivation.

Finally, intelligence alone does not build emotional resilience. Smart students may appear confident on the outside while struggling internally with uncertainty, setbacks, or social challenges. Resilience is learned through engagement with difficulty, not avoidance. Without it, intelligence can feel like a brittle asset. Growth requires pairing cognitive ability with emotional and behavioral skills. When this combination develops, intelligence transforms from a static label into a dynamic advantage.

Smart students often underestimate the importance of incremental progress. Because they grasp concepts quickly, they may expect immediate results in every endeavor. When outcomes lag, frustration sets in and motivation declines. The contrast between expected and actual performance can feel like failure. Yet real growth is cumulative, built through repetition and refinement. Recognizing the value of small steps is a turning point in breaking the stuck cycle.

Another subtle trap is the reliance on natural ability rather than learned strategies. Early success reinforces the idea that effort is optional. Students may skip learning habits that could sustain long-term achievement. This reliance becomes evident when challenges increase in complexity. Without intentional strategy, intelligence cannot carry them forward indefinitely. Developing systems for learning and execution is critical.

The culture of "smartness" also discourages vulnerability. Admitting confusion or asking for help can feel like risking identity. Smart students often internalize the belief that they must appear capable at all times. This mindset prevents collaboration and slows skill acquisition. Growth requires moments of exposure and humility. By allowing themselves to be seen as learners, students gain access to deeper understanding.

Expectations attached to intelligence can also distort priorities. Students may focus on tasks that highlight their ability rather than those that advance their development. High grades, recognition, or praise may overshadow personal growth. This misalignment reinforces stagnation. Refocusing on mastery over image allows intelligence to serve growth rather than status.

Finally, intelligence must be paired with action to remain effective. Thinking alone does not produce results or insights. Execution generates feedback, reveals blind spots, and builds confidence. Students who act, fail, and adapt develop momentum that pure intellect cannot achieve. Recognizing this principle transforms intelligence from a static gift into a dynamic tool for growth.

Smart students often confuse potential with progress. Being intelligent feels like having a head start, but without consistent action, potential remains unrealized. The danger is believing that natural ability guarantees results without effort. Over time, this creates frustration and self-doubt. Recognizing that ability is a foundation rather than a finished product is essential. True growth comes from building on that foundation deliberately and persistently.

Early praise for intelligence can also unintentionally create fear of failure. Students become conditioned to expect success without struggle. When challenges arise, they may interpret difficulty as evidence of inadequacy. This misinterpretation discourages experimentation and risk-taking. Over time, the student may avoid challenges altogether. Learning to separate temporary setbacks from identity is a key step in overcoming stagnation.

Another hidden consequence is a narrow focus on what intelligence measures. Academic tests, quizzes, and standardized assessments reward memorization and quick reasoning. Real-world problems demand persistence, creativity, and interpersonal skills—areas often neglected in early praise for intelligence. Focusing only on measurable outcomes limits growth potential. Broadening the scope of what counts as achievement is necessary for momentum.

Smart students also face the challenge of internalized comparison. Being labeled "smart" early fosters a mindset of competition, even with oneself. Every task becomes a test of ability rather than a learning opportunity. This comparison increases stress and decreases enjoyment of the process. Shifting focus from outperforming others to mastering personal goals transforms experience. Growth becomes self-directed rather than externally dictated.

Finally, the most critical insight is that intelligence alone cannot produce fulfillment or sustainable success. It is a tool, not a guarantee. Without effort, resilience, and emotional intelligence, ability alone can feel empty. When students combine intelligence with curiosity, persistence, and action, they unlock true potential. The feeling of being stuck begins to dissolve when intellect becomes partnered with deliberate growth.

QUOTES:

"The mind is not a vessel to be filled but a fire to be kindled." — Plutarch

Chapter .2

When Talent Stops Feeling Like Progress

Talent can feel like a natural advantage, but early ability often masks the work required for long-term growth. Many students breeze through school and early challenges without developing strategies for sustained effort. Initially, success comes easily, reinforcing the belief that progress will always feel effortless. When obstacles become more complex, the same talent that once provided confidence can feel insufficient. The shift from easy wins to sustained effort is jarring. Recognizing this change is the first step in transforming talent into real achievement.

Early achievement can create a false sense of momentum. When tasks are simple, speed and accuracy are rewarded over persistence. Students may assume that future challenges will also be conquered effortlessly. This assumption sets up a mismatch between expectation and reality. The first time progress slows, frustration and doubt can take hold. The gap between talent and sustained accomplishment becomes painfully visible.

Another factor is that talent often encourages shortcuts. Smart students may rely on intuition or quick comprehension rather than structured practice. When early success validates this approach, it becomes reinforced. Over time, these shortcuts fail to produce growth in more complex or unfamiliar situations. Students may feel stuck because their habitual strategies no longer suffice. Talent alone cannot compensate for a lack of depth or endurance.

Talent can also discourage risk-taking. When students succeed easily, they may prefer tasks where they are already strong. New challenges feel threatening because they might expose areas of weakness. This avoidance limits growth and reduces opportunities for learning. Progress feels stalled, not because of ability, but because of hesitation. Growth requires stepping into discomfort repeatedly.

Finally, talent can obscure the difference between understanding and application. A student may grasp concepts quickly but struggle to implement them in practice. This disconnect can be frustrating and confusing. Talent accelerates comprehension but does not automatically create competence. Progress slows when execution lags behind understanding. Learning to translate ability into consistent results is a critical skill for long-term success.

Talent is a gift, but without intentional effort, it can feel like a trap. Many students interpret early ease as a promise of continued success. When that promise is broken by more complex challenges, it can lead to self-doubt and stagnation. Recognizing that talent must be paired with skill-building, persistence, and experimentation transforms potential into measurable progress. Understanding this principle is key to moving beyond the stuck feeling.

The illusion of effortless progress is compounded by external expectations. Teachers, parents, and peers may assume that talent guarantees outcomes. Praise often focuses on results rather than effort, reinforcing a performance-oriented mindset. Students may internalize the belief that slowing down indicates failure. This mindset creates invisible pressure that makes growth feel like an uphill battle. Reframing success as consistent effort rather than immediate results is essential.

Students with early talent may also underestimate the importance of deliberate practice. Mastery requires more than natural ability; it demands structured effort and repetition. Skipping practice may have worked in simpler contexts, but complex tasks reveal gaps.

Talent may allow initial progress, but refinement comes only through effort. Recognizing the limits of natural ability fosters humility and resilience. Growth becomes intentional rather than accidental.

Relying solely on talent can also create unrealistic benchmarks. Students compare themselves to peers, historical standards, or imagined ideals. This comparison often leads to frustration when progress is slower than expected. Talent may have previously masked these gaps, but real-world challenges expose them. Learning to set personal, achievable benchmarks protects motivation. Progress becomes a measure of effort and growth rather than comparison.

Early talent can create a false sense of self-sufficiency. Students who grasp concepts quickly may feel they do not need guidance or feedback. This independence is reinforced by initial successes, giving the impression that help is unnecessary. Over time, this mindset can isolate learners from valuable support. Opportunities to correct mistakes or refine strategies are missed. True progress often depends on knowing when to seek assistance and applying it effectively.

Talent can also heighten the fear of plateauing. Students accustomed to rapid progress may become anxious when improvement slows. They may interpret stagnation as a personal failure rather than a natural part of growth. This perception can reduce risk-taking and experimentation. Ironically, the very talent that enabled early achievement now amplifies fear of slowing down. Learning to accept and work through plateaus is essential for long-term development.

Another challenge is the reliance on intuitive skill over structured strategy. Early talent allows students to solve problems without systematic approaches. While this feels efficient, it creates a vulnerability when problems become complex. Intuition alone cannot handle unfamiliar challenges consistently. Developing deliberate strategies is necessary to sustain high performance. The combination of talent and strategy produces reliable and scalable success.

Talent can also encourage avoidance of effortful practice. When natural ability produces results quickly, students may undervalue repetition, reflection, and refinement. Skipping these steps seems efficient but limits depth and resilience. Over time, gaps accumulate and manifest when tasks increase in difficulty. Students may feel stuck because their talent cannot bridge unpracticed skills. Recognizing that effort complements ability transforms early promise into enduring capability.

Talent can create a sense of entitlement to success. Students accustomed to early wins may assume that their abilities alone guarantee positive outcomes. When challenges become more complex, this expectation is shattered, leading to frustration and self-doubt. They may resist putting in extra effort because it feels unnecessary or unfair. This mindset can slow progress and reinforce a stuck feeling. Recognizing that talent must be paired with persistence is a turning point in personal development.

Another hidden effect of early talent is the overemphasis on outcomes rather than learning. Students may focus on grades, awards, or recognition, mistaking them for proof of growth. This focus can distract from the process and reduce engagement with challenges that actually build skill. Long-term improvement requires prioritizing mastery over accolades. When students shift attention from results to development, they regain momentum. Talent then becomes a tool, not a crutch.

Talent can also intensify the fear of failure. High-ability students often define themselves by past successes. Each new challenge carries the risk of undermining that identity. This fear can cause avoidance of tasks that are difficult, uncertain, or unfamiliar. Ironically, avoiding challenges slows progress and limits the very growth their talent could support. Learning to view failure as feedback rather than threat is critical for sustained advancement.

Students with talent may also struggle to maintain curiosity. When things come easily, the drive to explore deeply can diminish. The thrill of discovery may be replaced with complacency, relying on skill rather than curiosity. Over time, learning becomes about maintaining competence instead of expanding horizons. Reconnecting with genuine curiosity reignites motivation and opens pathways for continued progress. Talent is most powerful when paired with ongoing exploration.

Over time, talent can create an uneven relationship with effort. Students who succeed easily may see hard work as optional or even unnecessary. When challenges increase in difficulty, this habit becomes a liability rather than an advantage. The gap between talent and skill widens, producing frustration and self-doubt. Learning to value effort as much as ability is essential for converting potential into real progress. Consistent action transforms talent into durable competence.

Talent can also create blind spots in self-awareness. Students accustomed to early success may fail to notice weaknesses until they become significant obstacles. Without reflection and feedback, these gaps go unaddressed. When confronted with challenges, students may feel unprepared despite natural ability. Regular evaluation of strengths and weaknesses allows talent to be applied strategically. Self-awareness ensures that potential is guided toward meaningful growth.

Another challenge is that talent may foster impatience. Students who grasp concepts quickly may expect the same pace in all areas of life. When progress slows, frustration and discouragement arise. This impatience can lead to quitting or switching paths prematurely. Sustainable growth requires tolerance for incremental improvement and delayed gratification. Developing patience transforms initial talent into long-term achievement.

Talent can sometimes foster a sense of isolation. Being ahead of peers in certain areas can make collaboration or group learning feel unnecessary or uncomfortable. Students may avoid teamwork, believing their ability is sufficient alone. This isolation limits exposure to new perspectives, feedback, and opportunities for improvement. True mastery combines talent with collaboration and learning from others. Social engagement strengthens the application of ability.

Talent often creates the illusion that success should feel easy, but real growth requires persistence. Smart students may become frustrated when challenges demand sustained effort or creativity. Early wins can mask the skills needed for long-term achievement, leaving students unprepared for complex tasks. When progress slows, motivation may falter, and the sense of being "stuck" intensifies. Recognizing that difficulty is a natural part of growth helps reframe obstacles as opportunities. Talent becomes most valuable when paired with effort and resilience.

Another barrier is the tendency to rely on shortcuts. Early success encourages using intuition and speed rather than developing deep understanding or structured approaches. These shortcuts work in simple contexts but fail in unfamiliar or complex situations. Students may feel trapped because their habitual methods no longer produce results. Learning to embrace deliberate practice bridges the gap between talent and sustainable progress. Gradually, the feeling of being stuck transforms into forward motion.

The pressure of expectation also compounds the challenge. When peers, teachers, and family assume success is automatic, students internalize a high standard for themselves. This pressure can lead to fear of failure or hesitation to explore new paths. Early talent creates the perception that any setback is a personal flaw. Understanding that difficulty is normal for growth reduces anxiety and opens space for experimentation. Students can begin to engage fully with learning rather than guarding their image.

Talent alone is insufficient for complex problem-solving. Real-world challenges often require persistence, adaptability, and collaboration—qualities that are not automatically developed through early success. Students may feel confident in their knowledge yet struggle to apply it effectively. By recognizing the limits of innate ability, they can begin to build complementary skills intentionally. Combining talent with strategy, effort, and reflection creates a reliable path toward meaningful achievement.

Finally, the most important lesson is that talent must evolve into skill through sustained action. Early ability gives a head start, but it does not guarantee mastery or fulfillment. Students must learn to embrace effort, seek feedback, and confront challenges directly. Each small, deliberate step transforms latent potential into tangible progress. When talent is paired with persistence, strategy, and self-awareness, feeling stuck becomes a temporary phase rather than a permanent state. Momentum replaces stagnation, and potential becomes realized achievement.

Chapter .3

The Hidden Cost of Being "The Smart One"

Being labeled "the smart one" can feel like a badge of honor, but it comes with invisible burdens. From early on, teachers, peers, and family members place expectations on students based on their perceived intelligence. Every success is noticed, and every struggle feels magnified. This label shapes identity, influencing choices and behavior in subtle ways. Students may begin to define themselves entirely by their ability rather than by curiosity, effort, or creativity. Over time, this identity can limit growth and increase anxiety about making mistakes.

One major cost is the avoidance of risk. When being smart becomes central to identity, challenges that might expose weakness are intimidating. Students may stick to familiar tasks where they can excel effortlessly. This safe zone feels comfortable but restricts learning and skill development. Avoidance becomes a self-protective mechanism, but it also reinforces stagnation. Growth requires stepping into uncertainty despite fear of appearing less capable.

The label also encourages a focus on performance over learning. Smart students may prioritize outcomes that validate intelligence—grades, awards, recognition over genuine engagement with challenging material. This creates a narrow definition of success that limits exploration and creativity. Mistakes are seen as threats rather than feedback, slowing progress. When performance overshadows learning, students become more concerned with maintaining an image than developing ability.

Social dynamics add another layer of complexity. Peers may admire, envy, or even resent "the smart one." Relationships can be influenced by status rather than genuine connection, creating isolation. Students may hesitate to ask for help to avoid appearing less capable. This isolation reinforces the pressure to perform perfectly. Emotional strain accumulates quietly, even as outward success continues.

The "smart one" label can also create unrealistic internal expectations. Students may feel they must always understand things quickly or excel without visible effort. This expectation turns every setback into a perceived personal failure. Struggling with new or complex material may provoke anxiety rather than curiosity. Over time, these internal pressures compound, making growth feel risky or uncomfortable. Learning to separate ability from identity allows students to engage with challenges more freely.

Another hidden cost is the fear of judgment. Students labeled as smart often assume that asking questions or making mistakes will disappoint others. This fear can silence curiosity and limit engagement. Opportunities to clarify understanding, seek mentorship, or experiment may be avoided. The desire to maintain an image of intelligence outweighs the benefits of active learning. By reframing mistakes as evidence of growth rather than weakness, students regain confidence. The shift transforms the learning process from performance-based to development-focused.

Being "the smart one" can also lead to comparison traps. Students may constantly measure themselves against peers or past achievements. This competitive mindset fosters anxiety and distracts from personal growth. Instead of focusing on skill development, students focus on validation. Over time, even meaningful accomplishments can feel inadequate. Cultivating a mindset of self-referenced progress rather than external comparison mitigates this hidden cost.

The label can further encourage perfectionism disguised as high standards. Students may feel they are expected to excel in every subject or project, leaving little room for experimentation. Effort that falls short of excellence may feel like failure. This mindset discourages risk-taking and slows learning. Recognizing that imperfection is a natural part of mastery allows students to engage with challenges without fear. Perfectionism shifts from a barrier into a framework for deliberate improvement.

The "smart one" identity can create a dependency on external validation. Praise and recognition reinforce the sense of self-worth tied to intelligence. Students may begin to measure their value by how others perceive their ability. This reliance makes it difficult to take on challenges without immediate feedback or approval. Over time, intrinsic motivation can weaken, leaving effort contingent on recognition. Learning to cultivate internal validation strengthens resilience and fosters autonomous growth.

Another hidden cost is the avoidance of failure at all costs. Students may interpret setbacks as proof of inadequacy rather than opportunities to learn. This mindset encourages sticking to what is easy or familiar. Over time, avoidance reduces exposure to experiences that build new skills and adaptability. True growth requires embracing situations where mistakes are likely and learning is inevitable. By redefining failure as feedback, students reclaim the freedom to experiment and improve.

The pressure to maintain a "smart" image can also interfere with collaboration. Students may hesitate to ask questions, admit gaps, or engage in brainstorming for fear of appearing less capable. This behavior limits the exchange of ideas and reduces learning opportunities. Peer relationships may become transactional, based on performance rather than curiosity. Recognizing the value of vulnerability and collaboration allows intelligence to be applied more effectively. Growth is amplified when students combine ability with openness.

Being labeled smart can also create a sense of rigid identity. Students may feel trapped by the expectation that they must always be the fastest, brightest, or most capable in every situation. This rigidity discourages exploration of new interests or unconventional paths. Curiosity is constrained by the need to perform within a narrow definition of intelligence. When students allow identity to be flexible, they regain the ability to experiment, take risks, and grow in diverse ways. Adaptability becomes a key advantage.

The "smart one" identity often encourages over-planning and hesitation. Students may spend excessive time analyzing tasks to ensure perfect execution. This behavior can slow progress and reduce opportunities to learn from real-world experience. Overthinking becomes a subtle form of self-protection, avoiding mistakes at all costs. The irony is that intelligence, which enables strategic thinking, can also reinforce paralysis. Learning to balance planning with action allows students to transform caution into momentum.

Another hidden cost is the misalignment between passion and perceived ability. Students may pursue subjects where they are expected to excel, even if these areas do not inspire them. External expectations can overshadow intrinsic interests. Over time, engagement diminishes, and motivation wanes despite high performance. Recognizing personal passions and integrating them with ability reignites curiosity and commitment. Aligning talent with genuine interest enhances both fulfillment and achievement.

Being labeled "the smart one" can also generate fear of social judgment. Students may hesitate to share ideas or take unconventional approaches, fearing peer criticism or failure. Conformity often replaces innovation in order to maintain the image of competence. Over time, this can stifle creativity and reduce problem-solving opportunities. Embracing constructive risk and accepting potential judgment allows intelligence to flourish. Freedom to experiment fosters confidence and deeper growth.

The identity can additionally create an internalized "comparison trap." Students often measure themselves against past successes or the achievements of peers. This comparison generates stress, undermines satisfaction, and distracts from self-directed learning.

Progress slows when focus is on ranking rather than developing. Shifting attention toward personal growth and incremental improvement mitigates the negative effects of comparison. Internal benchmarks encourage consistent, meaningful progress.

The "smart one" label can also make students resistant to feedback. When identity is tied to intelligence, criticism can feel personal rather than constructive. Students may dismiss advice or avoid situations where they could be evaluated. Over time, this limits growth and reinforces a sense of stagnation. Learning to separate feedback from self-worth allows intelligence to be applied more effectively. Openness to critique transforms challenges into opportunities for improvement.

Another hidden cost is over-reliance on intellectual problem-solving. Students may default to thinking their way through every challenge, underestimating the value of experimentation, practice, or collaboration. While intelligence provides insight, it does not guarantee real-world success on its own. Students may feel frustrated when thinking alone does not produce results. Integrating action, reflection, and teamwork amplifies the impact of talent. The combination of intellect and applied effort produces meaningful progress.

The "smart one" identity can foster a perfectionist mindset. Students may avoid tasks where they might struggle, fearing mistakes would tarnish their image. This avoidance prevents engagement with learning opportunities that build resilience and adaptability. Over time, perfectionism restricts exploration and slows development. By reframing imperfection as a necessary part of mastery, students can embrace challenges more fully. Growth occurs when mistakes are seen as tools rather than threats.

Social isolation is another subtle consequence. Peers may defer to the "smart one" or avoid collaboration, assuming answers are already known. This dynamic can reduce opportunities for discussion, feedback, and shared problem-solving. Students may feel pressure to maintain competence rather than connect authentically. Building supportive relationships and seeking collaborative learning experiences counteracts this isolation. Connection enhances both learning and emotional well-being.

Finally, the hidden cost of being “the smart one” is the restriction it places on curiosity and risk-taking. Students may pursue predictable paths where success is likely, avoiding new or challenging experiences. Over time, this narrows potential and limits personal growth. Recognizing that intelligence is a tool, not a definition of identity, empowers students to explore, fail, and innovate. Freedom from the label allows them to take risks, embrace challenges, and transform talent into sustained achievement.

Chapter .4

Why Effort Feels Like Failure to Smart Students

For many smart students, effort is uncomfortable because it challenges a long-held identity. When intelligence has been associated with ease, working hard can feel unnatural or even shameful. Struggling with tasks may trigger self-doubt, as if effort itself signals inadequacy. This mindset creates a paradox: the more capable a student is, the more effort feels like evidence of failure. Over time, this association discourages persistence and experimentation. Understanding that effort is a normal and necessary part of growth is essential for overcoming this barrier.

Early success often conditions students to expect quick results. Tasks that once came easily reinforce the idea that intelligence should automatically translate to outcomes. When later challenges require sustained work, students may misinterpret the effort as a lack of ability. Frustration mounts because progress feels slower than expected. This mismatch between expectation and reality contributes to the "stuck" feeling. Learning to recalibrate expectations allows effort to be seen as part of the process rather than a sign of weakness.

Effort can also feel like failure because smart students are highly attuned to external evaluation. Grades, praise, and recognition become benchmarks of competence. When outcomes do not immediately reflect effort, students may perceive themselves as failing. This external feedback loop reinforces anxiety and self-judgment. By focusing on personal growth rather than external validation, effort can be reframed as constructive action. Progress is measured by learning and persistence, not immediate results.

Perfectionism compounds the problem. Students may associate effort with imperfection, assuming that requiring practice signals deficiency. Tasks that take longer than expected may feel threatening to identity. This mindset discourages experimentation and encourages avoidance of challenging tasks. Overcoming this requires redefining effort as a pathway to mastery rather than a reflection of current ability. Embracing imperfection allows students to engage fully without fear.

Effort feels like failure when students have internalized the idea that smart people should succeed effortlessly. This expectation turns normal challenges into a source of anxiety. Tasks that require sustained focus or repeated practice feel like proof of inadequacy rather than opportunity. As a result, students may avoid difficult work or give up prematurely. Over time, this pattern reinforces the stuck feeling. Recognizing that effort is a sign of engagement, not incompetence, is key to breaking the cycle.

Another reason effort feels like failure is that students often compare themselves to peers who appear to excel naturally. Observing others succeed without visible struggle can create the illusion that they are lagging. This comparison amplifies self-criticism and undermines confidence. The gap between perceived ability and actual progress seems larger than it truly is. Shifting focus from comparison to personal improvement helps students appreciate the value of sustained work. Progress becomes measurable by growth rather than by others' performance.

Smart students may also interpret mistakes as personal flaws. Every error can feel magnified because of the expectation of innate competence. Struggling with tasks is seen not as a learning opportunity but as confirmation of inadequacy. This mindset discourages experimentation and slows development. Recognizing that mistakes are natural and instructive transforms the way effort is perceived. Effort becomes a pathway to understanding rather than a reflection of identity.

Perfectionism reinforces the discomfort of effort. Students may believe that if something is done well, it should require minimal time or energy. When a task takes effort, they may assume they are failing. This belief discourages persistence and encourages avoidance. By reframing effort as essential to mastery, students can embrace the process rather than fearing it. Growth is then measured by engagement and learning, not by immediate ease.

Effort can also feel like failure because it makes progress visible in small, incremental steps rather than dramatic leaps. Smart students who are used to quick wins may become frustrated when improvement comes slowly. The lack of immediate, noticeable results can be discouraging. They may interpret gradual learning as stagnation or inadequacy. This perception undermines motivation and reinforces the belief that effort is failing. Learning to value incremental progress helps students recognize that effort is cumulative, not instant.

Another reason is the overemphasis on outcomes rather than process. When results are prioritized above learning, the time and energy invested can feel wasted if success is not immediate. Students may believe that struggling to achieve a goal reflects poorly on their ability. This mindset prevents them from appreciating the growth happening beneath the surface. By focusing on what is learned and skills developed, effort becomes meaningful rather than threatening. Reframing effort as a tool for growth rather than a test of ability shifts the entire learning experience.

Effort may feel like failure when students rely on natural ability to carry them forward. When intelligence previously produced fast success, the need to work harder can feel unfamiliar and uncomfortable. Tasks requiring persistence, strategy, or repeated practice may trigger self-doubt. Students may interpret the need for effort as a signal that they are no longer smart. This misinterpretation creates anxiety and avoidance. Accepting that all meaningful achievement requires effort normalizes struggle as part of growth.

Social comparison further amplifies the problem. Seeing peers succeed with apparent ease can make students question their own abilities. They may overlook the effort others have invested behind the scenes. This distorted perception makes personal effort feel insufficient or wrong. By recognizing that everyone struggles in some way, students can reduce anxiety and reframe effort as necessary. Social perspective allows intelligence to work in tandem with persistence.

Effort feels like failure when students set unrealistic standards for themselves. Being smart often creates the expectation that tasks should be mastered quickly and flawlessly. When outcomes do not meet these self-imposed standards, students may interpret the effort they put in as insufficient or wrong. This mindset encourages discouragement and avoidance rather than persistence. Over time, it erodes confidence and reduces willingness to take on challenges. Learning to set realistic goals and value the process over perfection transforms effort into a source of growth.

Another reason effort feels threatening is the internalized pressure to prove intelligence. Students may worry that visible struggle will undermine how others perceive them. They may hide effort, avoid seeking help, or focus only on tasks they can complete easily. This self-protective behavior limits learning and reinforces the idea that working hard is a weakness. Recognizing that effort is a natural part of achievement helps students reframe struggle as strength. Embracing effort allows intelligence to be fully applied without fear.

The discomfort of effort is often compounded by impatience. Students accustomed to quick success may expect immediate results from new challenges. When progress is slow, they may feel frustrated or demotivated. This impatience can discourage repeated practice or sustained engagement. Learning to tolerate gradual improvement and delayed gratification is crucial for turning effort into mastery. Patience transforms effort from a source of anxiety into a pathway for growth.

Effort may also feel like failure when students focus solely on outcomes rather than learning. Achievements are celebrated, but the process that leads there is overlooked. As a result, the time and energy invested in developing skills can feel wasted. Recognizing that every action contributes to competence shifts perspective. Effort becomes meaningful, measurable, and rewarding even before final results are achieved. This shift fosters motivation and reinforces resilience.

Effort can feel like failure when students confuse hard work with inadequacy. Because intelligence was rewarded for ease early on, any situation requiring sustained effort may trigger self-doubt. Students may believe that struggling signals a lack of ability rather than an opportunity to grow. This perception discourages persistence and encourages avoidance of challenging tasks. Reframing effort as a normal and necessary component of learning shifts its meaning entirely. Growth occurs when effort is understood as a tool, not a judgment.

Another challenge is the expectation of immediate mastery. Smart students are often praised for quick comprehension, which sets an internal standard of instant success. When learning takes time, patience is tested. Students may perceive effort as a sign that they are falling behind, even when progress is occurring steadily. Recognizing that mastery develops incrementally helps normalize effort. Gradual improvement becomes evidence of engagement rather than failure.

Effort can also feel like failure when feedback is inconsistent or delayed. Students may invest time and energy into tasks without clear indicators of progress. Without immediate reinforcement, hard work can seem futile. This perception reinforces the idea that effort is ineffective. By tracking incremental gains and focusing on personal benchmarks, students learn to appreciate their efforts. Feedback becomes a tool for growth rather than a verdict on ability.

Perfectionism further complicates the experience of effort. The desire to perform flawlessly can make every challenge feel high-stakes. Tasks that require practice and iteration may feel threatening to identity. Students may avoid these tasks entirely or engage half-heartedly. Embracing imperfection as an essential part of mastery transforms effort into a learning opportunity. Each attempt becomes a step toward skill rather than a measure of worth.

Finally, the most important realization is that effort is not failure—it is the vehicle for progress. Smart students often confuse the necessity of hard work with evidence of inadequacy. When this mindset is shifted, effort becomes empowering rather than discouraging. Challenges, struggle, and repeated practice are no longer signs of limitation but markers of growth. Students who embrace effort gain confidence, build resilience, and unlock the full potential of their intelligence. Persistence transforms ability into achievement.

Chapter .5

How School Rewards the Wrong Kind of Thinking

Traditional schooling often emphasizes speed, compliance, and correctness over curiosity, critical thinking, and creativity. From a young age, students are trained to follow rules, memorize information, and produce answers quickly. This system rewards conformity rather than experimentation and values right answers over learning how to think. Smart students may excel within these boundaries, but their development in adaptability, problem-solving, and resilience is often neglected. The habits cultivated by schooling can make real-world challenges feel foreign and intimidating. Understanding this mismatch is essential for turning intelligence into meaningful growth.

One consequence of this system is that deep thinking is often discouraged. Lessons are structured to move quickly, and assessments focus on outcomes rather than process. Students who prefer to analyze, reflect, or explore multiple approaches may feel pressured to conform to faster, shallower methods. The focus on speed over depth reinforces the perception that intelligence is a matter of efficiency rather than insight. Over time, this creates a habit of prioritizing quick performance instead of thoughtful problem-solving. Developing the ability to think deeply requires unlearning these school-conditioned patterns.

Another effect is the overemphasis on grades and standardized measures. Smart students may equate their worth with performance on tests, quizzes, or assignments. Learning becomes a tool for validation rather than discovery. As a result, curiosity is replaced by calculation, and genuine engagement diminishes. The focus on outcomes over process also discourages risk-taking, since errors may directly affect grades. Breaking free from this mindset requires valuing effort, exploration, and experimentation over external approval.

School systems also reward compliance more than initiative. Following instructions correctly is often valued more than questioning, experimenting, or innovating. Students may be discouraged from challenging assumptions or proposing new ideas. Over time, this trains smart students to defer judgment and rely on authority rather than their own reasoning. Developing independent thinking involves cultivating the courage to act thoughtfully outside the confines of predetermined rules. True learning occurs when students move beyond compliance and actively explore possibilities.

QUOTES:

"Education is not the filling of a pail, but the lighting of a fire." — William Butler Yeats

CHAPTER .6

THE CONFIDENCE–COMPETENCE GAP

Smart students often experience a disconnect between how capable they feel and how effective they actually are. Confidence may be based on prior successes rather than demonstrated skill in new or complex contexts. This gap can lead to overestimating ability in unfamiliar situations or underestimating it when challenges arise. Students may hesitate to take action, fearing that effort will expose inadequacy. Over time, this gap creates frustration, stagnation, and self-doubt. Bridging the divide between confidence and competence is essential for turning potential into real-world results.

Early achievements can inflate confidence while leaving competence underdeveloped. When tasks were easy, students learned that natural ability was sufficient. They may not have practiced persistence, strategy, or problem-solving in depth. When new challenges arise, confidence remains high but is not backed by the necessary skills. This mismatch produces anxiety, avoidance, or poor decision-making. Recognizing the limits of confidence helps students approach challenges more deliberately.

The gap is further widened by social perception. Praise, awards, and recognition can reinforce confidence without providing evidence of skill mastery. Students may assume competence because others believe in it, masking areas that require growth. When real-world demands expose these gaps, the sense of being "stuck" intensifies. Developing competence through deliberate practice ensures that confidence is rooted in tangible ability. Over time, alignment between belief and skill fosters both effectiveness and resilience.

Another contributor is the habit of overthinking rather than acting. Students may feel capable in theory but fail to translate ideas into practice. The act of doing exposes weaknesses and creates feedback for improvement. Without action, confidence remains abstract, and competence does not grow. Learning to test knowledge through real-world application bridges the gap. Execution transforms potential into measurable skill.

Confidence without competence often leads to miscalculated risks. Smart students may feel capable of handling challenges they have little experience with, assuming prior success will carry them forward. When results fall short of expectations, the disconnect between confidence and skill becomes painfully clear. This can result in embarrassment, self-doubt, or avoidance of future challenges. Over time, repeated exposure to this gap reinforces the perception that confidence is unreliable. Bridging the gap requires intentionally developing skills alongside self-assuredness.

Another challenge is that overconfidence can mask the need for growth. Students may skip practice, avoid feedback, or neglect preparation, believing their natural ability is sufficient. This mindset prevents competence from catching up to confidence. The mismatch between belief and ability creates frustration when tasks become more difficult. Cultivating humility alongside confidence encourages consistent effort and skill development. Awareness of limits allows students to act effectively without overestimating their capacity.

The confidence–competence gap can also distort decision-making. Students may take action prematurely or overcommit because they feel capable, but their skills are not yet fully developed. Conversely, they may hesitate on tasks that require effort or uncertainty, doubting their ability despite prior successes. Both overextension and underaction stem from the same gap. Learning to assess challenges realistically and act deliberately helps align confidence with competence. Balanced judgment fosters more consistent success and reduces anxiety.

External reinforcement can exacerbate the gap. Praise, accolades, and recognition often reward outcomes without addressing the underlying abilities needed for consistent performance. Students may feel accomplished based on appearance rather than demonstrated skill. When new or ambiguous challenges appear, this overreliance on external validation leaves them unprepared. Developing competence through deliberate practice and feedback ensures that confidence is earned and reliable. This alignment strengthens self-trust and resilience.

Confidence can feel misleading when competence has gaps that go unnoticed. Smart students may excel in familiar areas, giving a sense of mastery, but falter when tasks demand new skills. This selective competence can create overconfidence in some situations and underconfidence in others. Students may misjudge challenges, either taking on too much or avoiding important opportunities. Recognizing areas that need growth is critical to bridging the confidence–competence gap. Targeted practice ensures that belief in one's abilities matches real-world capability.

The gap also affects problem-solving under pressure. Students with high confidence may rush decisions, assuming their intelligence alone will suffice. Conversely, low competence in untested areas can create hesitation or paralysis. Both extremes lead to suboptimal outcomes and reinforce the perception of being "stuck." Developing competence through deliberate, low-stakes practice builds reliability. Confidence becomes a stable foundation for action rather than a fragile assumption. The alignment of thought and skill allows for smoother navigation of complex tasks.

Peer perception can intensify the gap. Smart students are often expected to excel, which reinforces confidence without necessarily increasing competence. They may feel pressure to appear capable at all times, masking gaps in knowledge or skill. When challenges expose these gaps, self-doubt and stress increase. Seeking feedback and practicing humility help balance internal confidence with actual competence. Aligning perception with ability fosters both authenticity and effectiveness.

The gap can also distort learning habits. Students may avoid deep practice, thinking their intelligence alone will carry them through. They may focus on quick wins or easy tasks instead of deliberate effort. Over time, this limits skill development and reinforces the confidence–competence mismatch. Structured learning routines that prioritize mastery over performance help close the gap. Students learn to act with both belief and ability in alignment.

Overconfidence can create blind spots that mask areas in need of improvement. Smart students may assume they understand concepts fully when gaps remain hidden. This illusion prevents deliberate practice and slows skill development. When challenges arise that expose these gaps, students can feel frustrated or discouraged. Awareness of limitations is essential for aligning confidence with competence. Actively seeking feedback and testing assumptions helps bridge the gap between perceived and actual ability.

The gap often leads to inconsistent performance. In familiar situations, confidence aligns with competence, producing strong results. In new or complex contexts, confidence may exceed skill, leading to mistakes or setbacks. Conversely, students may underestimate their competence in areas they have less experience with, avoiding challenges unnecessarily. This inconsistency reinforces the feeling of being "stuck." Developing strategies to assess one's abilities realistically promotes steady growth and reliable performance.

Another subtle effect is that the confidence–competence gap can undermine resilience. Students who appear capable may struggle internally when effort does not yield immediate results. Over time, repeated misalignment between confidence and actual ability can erode self-trust. Learning to persist through setbacks while building competence strengthens both skills and confidence. Resilience grows when students understand that temporary struggles are part of meaningful progress. The ability to act despite uncertainty becomes a hallmark of adaptive intelligence.

The gap also influences risk-taking behavior. Students with inflated confidence may take on tasks beyond their current ability, leading to failure and discouragement. Those with deflated confidence may avoid opportunities they are fully capable of handling. Both tendencies limit growth and prevent momentum. Developing accurate self-assessment skills allows students to engage challenges at the right level. This balance ensures that confidence supports competence rather than undermines it.

The confidence–competence gap can also create hesitation in decision-making. Students may feel capable in theory but second-guess themselves in practice, worrying that mistakes will reveal inadequacy. This over-deliberation slows action and prevents skill from being tested and strengthened. Repeated hesitation reinforces self-doubt, widening the gap further. Smart students must learn to act despite uncertainty, using each experience as feedback for growth. Action becomes the bridge that converts confidence into validated competence.

Another effect is the tendency to overprepare. Students may spend excessive time studying, planning, or rehearsing because confidence feels insufficient to compensate for skill gaps. While preparation can be useful, overdoing it delays meaningful practice and real-world application. This cycle creates frustration, as students feel busy but unproductive. Learning to balance preparation with action ensures that competence develops alongside confidence. Practical experience validates ability and reinforces self-trust.

The gap can also distort how students perceive setbacks. When confident but underprepared, early failures may feel catastrophic rather than instructional. Conversely, those with low confidence may dismiss their successes, believing results are due to luck or external factors. Both perspectives prevent accurate assessment of skill and impede growth. Developing a reflective mindset allows students to extract lessons from every outcome. Over time, this reduces anxiety and aligns confidence with real competence.

Social pressures further amplify the confidence–competence mismatch. Peers, teachers, and parents often assume that smart students should perform perfectly. Praise reinforces confidence, while criticism can feel devastating. Students may hesitate to engage with challenging tasks for fear of exposing gaps. By cultivating self-awareness and internal validation, they learn to navigate expectations without losing momentum. Competence grows through practice, feedback, and self-directed effort rather than external judgment.

Chapter .7

When Overthinking Replaces Action

Smart students often get trapped in cycles of overthinking that prevent meaningful action. The ability to analyze deeply can become a liability when it delays execution. Students may evaluate every possible outcome, weigh pros and cons endlessly, and search for perfect information before acting. While careful thought has value, excessive analysis creates paralysis. Over time, opportunities are missed, and momentum stalls. Learning to balance thought with action is critical for turning intelligence into tangible results.

One reason overthinking dominates is the fear of making mistakes. Students may worry that a wrong choice will expose inadequacy or damage their reputation. This fear amplifies the tendency to ruminate and postpone decisions. Ironically, inaction often leads to worse outcomes than taking imperfect action. Recognizing that mistakes are feedback rather than failure shifts perspective. Acting despite uncertainty allows learning to occur in real time.

Overthinking can also arise from high expectations. Smart students may set standards that are impossible to meet, analyzing every task in search of perfection. This creates mental overload and reduces clarity. Tasks that could be completed with reasonable effort become sources of anxiety and procrastination. Developing the ability to prioritize, simplify, and focus on next steps reduces cognitive friction. Small, deliberate actions build momentum where analysis alone cannot.

Another contributor is the habit of seeking certainty. Students may believe that knowing every variable or outcome before acting is necessary. The illusion of certainty creates a waiting game that delays progress indefinitely. In reality, most meaningful learning occurs through experimentation and iteration. Accepting that uncertainty is part of growth allows action to proceed despite incomplete information. Movement generates clarity, revealing what works and what needs adjustment.

Overthinking often leads to unnecessary complexity. Smart students may break problems into countless components, analyzing each in isolation. While this can reveal details, it also creates mental clutter that hinders decision-making. Students may struggle to identify the most important actions and lose focus on achievable steps. Complexity becomes a barrier rather than a tool for insight. Simplifying problems and focusing on essential elements restores clarity and enables action.

Another consequence is delayed learning. When action is postponed in favor of extended analysis, students miss opportunities to gain real-world feedback. Without testing ideas through practice, understanding remains theoretical. This gap between thinking and doing can reinforce feelings of stagnation. Experiential learning is essential to bridge knowledge and competence. Taking deliberate action, even imperfectly, provides information that no amount of thinking can generate.

Overthinking is also fueled by fear of judgment. Students may anticipate how teachers, peers, or others will perceive their decisions. This concern can amplify indecision, as every option is scrutinized for potential criticism. The result is hesitation and avoidance of meaningful challenges. Learning to prioritize personal growth over external evaluation allows students to act confidently. Over time, repeated practice strengthens both competence and self-trust.

Perfectionism magnifies the cycle of overthinking. The desire to avoid mistakes or produce flawless outcomes can paralyze students before they even begin. Tasks that could be completed efficiently become sources of stress and procrastination. Reframing effort as iterative and progress as incremental reduces pressure. Students learn to value movement and experimentation over waiting for ideal conditions. Action, rather than endless contemplation, becomes the measure of growth.

Overthinking can also create a false sense of preparedness. Smart students may convince themselves that thinking longer equates to being ready. They may accumulate information, weigh scenarios, and plan endlessly, believing action must wait until everything is perfect. In reality, preparation without implementation rarely produces results. The gap between knowledge and execution grows wider with each delay. Learning to act while continuing to gather information bridges the divide between planning and doing.

Another issue is that overthinking can reinforce procrastination. Students may feel productive while analyzing, yet no tangible progress occurs. Time spent in rumination often replaces the energy needed for execution. Over time, tasks pile up, creating stress and diminishing motivation. Action, even imperfect action, is more productive than perfect analysis. Prioritizing doing over overthinking fosters forward momentum and tangible results.

Overthinking is often compounded by fear of irreversible consequences. Students may imagine worst-case scenarios for every decision, exaggerating potential risks. This mental rehearsal can make action feel too risky or intimidating. While considering consequences is valuable, obsessing over them prevents engagement. Learning to tolerate uncertainty and take calculated risks transforms analysis into actionable insight. Experience provides feedback that no amount of speculation can replicate.

The habit of overthinking can also hinder creativity. Students may overanalyze possibilities, evaluating each for feasibility, correctness, or risk. This approach can stifle intuition and spontaneous problem-solving. Creative solutions often emerge through experimentation rather than exhaustive planning. Taking imperfect action allows for discovery and innovation. By moving from analysis to application, students unlock their ability to generate new ideas.

Overthinking often prevents students from building momentum. Each task feels daunting because it is endlessly dissected in the mind before execution begins. The result is a cycle of inaction, where energy is spent thinking rather than doing. Momentum is essential for learning, as repeated action reinforces skill and confidence. When overthinking dominates, opportunities for growth are delayed or missed entirely. Breaking this cycle requires consciously shifting from analysis to small, deliberate steps.

Another consequence is mental fatigue. Constant rumination consumes cognitive resources, leaving students drained before they even act. Decision-making becomes exhausting, and motivation declines. Tasks that could be accomplished with reasonable effort feel overwhelming. Prioritizing actionable steps over exhaustive contemplation restores energy and clarity. The act of doing replenishes focus and reinforces competence. Gradual, consistent action counteracts the exhaustion caused by overthinking.

Overthinking can also create unrealistic expectations. Students may assume that if they plan thoroughly, outcomes should be perfect. When reality inevitably diverges from imagined scenarios, frustration arises. The perception that preparation should guarantee success reinforces the feeling of failure. Accepting that uncertainty is natural allows students to act without fear of imperfection. Experiencing outcomes firsthand provides practical feedback far more reliable than hypothetical planning.

Social and external pressures amplify the problem. Smart students may anticipate judgment from teachers, peers, or family, fueling excessive deliberation. The desire to appear competent often outweighs the willingness to experiment. This can prevent engagement with meaningful challenges and slow skill development. Shifting focus from external approval to personal growth encourages decisive action. Confidence is strengthened when students measure success by effort and learning rather than by others' perceptions.

The fear of inaction often fuels overthinking. Smart students may feel that every decision is critical, and any delay risks "wasting potential." This pressure makes starting seem daunting, and analysis becomes a substitute for action. Ironically, the longer they wait, the more opportunities slip by, reinforcing anxiety and indecision. Recognizing that momentum, not perfect timing, drives progress is crucial. Each small step forward reduces fear and builds confidence in the ability to act.

Overthinking can also create a false sense of control. Students may believe that analyzing every possible scenario protects them from mistakes. While planning has value, no amount of speculation can guarantee outcomes. The illusion of control can delay action indefinitely. Learning to embrace uncertainty allows students to take imperfect action with confidence. Action, combined with reflection, provides real-world feedback that planning alone cannot deliver.

The habit of overthinking can erode motivation over time. When mental energy is spent endlessly weighing options, the will to act diminishes. Tasks feel overwhelming before they are even attempted. Momentum stalls, and the perception of being "stuck" intensifies. Breaking the cycle requires conscious prioritization of action over analysis. Incremental steps create tangible results that reinforce both motivation and competence.

Social pressures compound the challenge. Smart students may fear judgment if their decisions lead to mistakes. This concern can amplify rumination and hesitation. Tasks that could be completed efficiently are delayed as students consider every possible consequence. Shifting focus from external validation to internal growth empowers students to act decisively. Each action builds competence, which in turn strengthens confidence and reduces fear.

Finally, the key to overcoming overthinking is iterative experimentation. Students must practice moving from thought to action, embracing mistakes as learning opportunities. Each small step produces evidence, feedback, and momentum, gradually replacing paralysis with progress. Over time, the habit of acting decisively strengthens both skill and self-trust. Thought becomes productive when paired with action, transforming intelligence into tangible achievement. Momentum replaces overanalysis, and smart students regain the ability to move forward with purpose.

CHAPTER .8

THE FEAR OF WASTING POTENTIAL

Smart students often experience a constant pressure to "live up" to their abilities. This fear can be paralyzing, as every choice feels like it could determine long-term success or failure. Opportunities may be avoided, decisions delayed, and risks minimized in an attempt to protect potential. Ironically, this caution often leads to underachievement and stagnation. Recognizing that potential is unlocked through action rather than perfection is key. Moving forward, even imperfectly, is what transforms ability into accomplishment.

The fear of wasting potential is closely tied to comparison with others. Students may measure themselves against peers who seem more accomplished or driven. Observing others' successes can create a sense of urgency and inadequacy, amplifying stress. This external comparison often distracts from personal goals and authentic growth. Reframing success around one's own path allows students to pursue meaningful action without the weight of constant benchmarking. Potential is best realized when aligned with intrinsic values rather than external standards.

Perfectionism intensifies this fear. Smart students may believe that anything less than extraordinary is a failure of potential. Ordinary steps feel insufficient, and progress can seem meaningless. This mindset discourages experimentation and incremental learning. By redefining progress as consistent effort and movement, students can act without fear of "wasting" their abilities. Even small actions accumulate into significant growth over time.

The fear of wasting potential also creates anxiety about long-term planning. Students may obsess over making the "right" decisions that maximize future outcomes. This focus can stall present action, leaving opportunities unexplored. In reality, potential is realized through repeated experimentation, reflection, and adaptation. Each decision, whether perfect or flawed, provides feedback and shapes growth. Trusting the process allows students to move forward while preserving their long-term aspirations.

The fear of wasting potential can also create indecision in choosing paths. Smart students may delay committing to projects, careers, or learning opportunities because they worry about "choosing wrong." Each option feels weighted with long-term consequences, and overthinking can dominate. This hesitation often leads to missed opportunities and regret, reinforcing the sense of being stuck. Recognizing that growth occurs through iterative choices allows students to act without fearing permanence. Movement and experimentation are the real keys to unlocking potential.

Another effect is chronic self-criticism. Students may constantly evaluate whether they are maximizing their abilities. Even meaningful accomplishments can feel inadequate because they fall short of imagined ideals. This persistent self-judgment creates stress and reduces enjoyment of progress. Shifting focus to measurable growth and consistent effort rather than perfection alleviates pressure. Celebrating small wins reinforces momentum and demonstrates that potential is being actively realized.

The fear of wasting potential can discourage risk-taking. Students may avoid challenges where failure is possible, fearing that mistakes will irreversibly harm their trajectory. Ironically, avoiding risk limits skill development, experience, and real progress. Smart students must learn that calculated risks are essential for growth. Each step into uncertainty provides valuable lessons that build competence. Embracing challenges gradually transforms fear into motivation.

Perfectionism compounds the anxiety associated with potential. Students may insist that every action, project, or choice must be executed flawlessly. This mindset makes even small tasks feel daunting and increases procrastination. By reframing mistakes as learning opportunities, students reduce the pressure to perform perfectly. Incremental improvement and experimentation become more valuable than immediate excellence. Effort and persistence, rather than perfection, become the markers of realized potential.

Finally, the fear of wasting potential can create a narrow definition of success. Students may believe that only certain achievements or accolades validate their ability. This limited perspective restricts exploration and curiosity, leaving potential untapped in other areas. Broadening the definition of success to include learning, growth, and experimentation encourages action. Recognizing that potential unfolds through sustained effort, not single achievements, releases students from paralysis. When fear is replaced by curiosity and engagement, smart students transform potential into meaningful, actionable progress.

Part II: Mental Traps That Keep Smart Students Stuck

CHAPTER .9

PERFECTIONISM DISGUISED AS HIGH STANDARDS

Perfectionism often masquerades as high standards, especially for smart students. They may insist that every task must be executed flawlessly, believing that anything less reflects a lack of ability. This mindset can appear disciplined and ambitious but often masks fear of failure. Tasks that could be completed efficiently become sources of stress and procrastination. Over time, this approach limits growth, creativity, and learning opportunities. Understanding the distinction between high standards and perfectionism is essential for breaking free from self-imposed barriers.

One way perfectionism manifests is through avoidance. Students may delay starting projects or exploring new areas because they fear they cannot do them perfectly. They may overanalyze or over prepare, waiting for ideal conditions that never arrive. This delay prevents skill development and reduces the ability to gain feedback from real-world application. High standards encourage refinement, while perfectionism creates paralysis. Learning to act despite imperfection allows students to make progress and build competence. Small, deliberate steps help bridge the gap between intention and achievement.

Perfectionism also fuels self-criticism. Every minor flaw or setback is amplified, interpreted as a reflection of personal inadequacy. Students may focus on what went wrong rather than what was learned. This mindset discourages experimentation and diminishes motivation. By reframing mistakes as feedback, effort becomes meaningful rather than threatening. Shifting focus from flawlessness to growth fosters resilience and continuous improvement.

Another consequence is the tendency to overcommit. Students may take on more than they can handle, believing they must excel in every domain. This spreads energy thin and reduces effectiveness. Tasks that could be completed well with focused effort become overwhelming. Prioritization and selective excellence help students maintain high standards without succumbing to perfectionism. Learning to allocate energy strategically allows for sustainable growth and achievement.

Perfectionism often prevents students from completing tasks. Because nothing feels "good enough," they may spend excessive time refining work or endlessly revising ideas. Projects that could be finished efficiently remain incomplete, reinforcing frustration and self-doubt. High standards encourage progress, but perfectionism traps students in cycles of revision and delay. Learning to set realistic goals and accept "good enough" allows action to take precedence over endless refinement. Completing tasks builds momentum, confidence, and tangible results.

Another effect is mental exhaustion. Constantly striving for flawlessness consumes cognitive and emotional energy. Students may feel drained even when little progress has been made. Over time, this leads to burnout, reducing motivation and engagement. Recognizing the cost of perfectionism helps students reallocate energy toward deliberate action and meaningful growth. Balanced effort ensures that intelligence is applied sustainably rather than being drained by unrealistic expectations.

Perfectionism also increases sensitivity to feedback. Students may interpret critique as proof of inadequacy rather than an opportunity to improve. This defensive reaction reduces openness to learning and collaboration. High standards, in contrast, allow constructive evaluation to guide progress without threatening self-worth. Embracing feedback as a tool for improvement transforms criticism into actionable insight. Over time, students learn to separate personal value from outcomes, reducing anxiety and fostering resilience.

The fear of imperfection can stifle creativity. Students may avoid exploring new ideas or taking risks, believing that failure or mistakes would compromise their perceived ability. Innovation requires experimentation, trial, and error—activities perfectionism discourages. By reframing errors as part of the learning process, students open themselves to discovery. Creativity thrives when action is prioritized over flawless execution. Momentum, curiosity, and adaptability replace hesitation and fear.

Perfectionism can also create a cycle of procrastination. Students may wait for the "perfect moment" to start a task or for conditions to feel ideal. This delay often prevents projects from ever being completed. Overthinking and excessive planning reinforce the belief that any action taken now would be inadequate. Learning to act in spite of imperfection breaks this cycle and builds confidence. Each completed task, no matter how small, demonstrates that progress matters more than flawlessness.

Another consequence is diminished self-esteem. Constantly chasing perfection can make students feel as if they are never doing enough. Achievements are undervalued, and even small mistakes feel catastrophic. Over time, this erodes confidence and increases anxiety. Recognizing that effort and growth are more important than perfection helps restore self-trust. Focusing on progress rather than flawlessness allows students to appreciate their accomplishments and build momentum.

Perfectionism also reduces resilience. When students equate mistakes with personal failure, any setback can feel devastating. This fear of failure discourages experimentation and risk-taking. Resilience grows when challenges are seen as learning opportunities rather than threats to identity. By embracing imperfection, students develop adaptability and problem-solving skills. Learning to recover and iterate strengthens both competence and confidence.

The pressure to appear flawless can also isolate smart students socially. Peers may perceive them as distant or unapproachable due to high standards. Students may avoid collaboration to prevent exposing weaknesses. This reduces opportunities for feedback, mentorship, and shared learning. By embracing imperfection, students open themselves to connection and collaboration. Interaction with others enhances learning, creativity, and personal growth.

Perfectionism often leads students to overcomplicate tasks. They may add unnecessary steps, revise repeatedly, or create elaborate systems in pursuit of flawlessness. This complexity slows progress and increases mental strain. High standards guide improvement, but perfectionism traps students in excessive refinement. Simplifying approaches allows focus on meaningful action and measurable results. By reducing unnecessary complexity, students can act more efficiently and build momentum.

Another effect is the fear of taking risks. Students may avoid challenging projects or new experiences because imperfection feels threatening. This avoidance limits opportunities for growth and skill development. In contrast, high standards encourage thoughtful effort and challenge without paralyzing action. Embracing risk, even when mistakes are possible, fosters learning and innovation. Students discover that progress often emerges from experimentation rather than flawless execution.

Perfectionism also affects time management. Tasks that could be completed in hours may stretch into days due to constant refinement and self-doubt. This inefficiency creates stress and reduces energy for other priorities. Learning to accept "good enough" and set realistic deadlines restores productivity. Time becomes a resource to advance growth rather than a source of anxiety. Strategic action ensures that effort produces tangible outcomes and sustained progress.

The habit of perfectionism can distort the perception of success. Students may dismiss accomplishments unless they meet impossible standards. This mindset reduces satisfaction and motivation, creating a cycle of dissatisfaction. High standards, on the other hand, allow achievement to be recognized while still encouraging improvement. By celebrating progress, students reinforce positive habits and maintain momentum. Satisfaction becomes a driver of continued effort rather than a rare, fleeting reward.

Perfectionism disguised as high standards can also erode intrinsic motivation. Students may focus on external validation—grades, praise, or recognition—rather than the joy of learning or discovery. When effort is driven by the need for flawlessness, curiosity and creativity take a backseat. Over time, this reduces engagement and enthusiasm for meaningful work. By shifting focus to personal growth and the process of learning, motivation becomes self-sustaining. Students learn to act because they value progress, not because they fear imperfection.

Another consequence is the tendency to overanalyze outcomes. Students may replay past tasks mentally, critiquing every perceived error. This reflection often leads to rumination rather than constructive learning. Perfectionism magnifies minor mistakes into major setbacks, reinforcing self-doubt. Developing a mindset that views errors as informative rather than catastrophic fosters resilience. Reflection becomes a tool for improvement, not a trap that prevents future action.

The habit of perfectionism can also create unrealistic comparisons with others. Students may feel that peers' achievements are effortless, intensifying pressure to perform flawlessly. This comparison can discourage collaboration and experimentation. In contrast, high standards encourage personal growth while allowing for imperfection and learning from others. Embracing collaboration and diverse perspectives enhances skill development and creative thinking. Smart students realize that growth is not a zero-sum competition.

Perfectionism further amplifies the fear of failure. Students may avoid challenges where success is not guaranteed, believing that mistakes will define their intelligence or worth. This avoidance limits learning, reduces experience, and hinders momentum. Accepting imperfection as part of the learning process allows students to take risks with confidence. Each attempt, regardless of outcome, builds competence and demonstrates that mistakes are not failures. Action becomes the measure of progress rather than flawless results.

Finally, overcoming perfectionism requires a deliberate shift in mindset and habits. Students must act consistently, embrace mistakes as feedback, and prioritize progress over flawlessness. High standards guide effort without paralyzing action, while flexibility allows for adaptation and learning. Each completed task reinforces competence, builds confidence, and reduces anxiety. By distinguishing between high standards and perfectionism, smart students unlock their potential and transform intelligence into sustained achievement.

QUOTES:

"Perfection is the enemy of progress." — Winston Churchill

CHAPTER .10

THE ILLUSION OF ONE PERFECT PATH

Smart students often believe that there is a single correct path to success, and this belief can be paralyzing. They may obsess over choosing the "right" college, career, or project, fearing that any deviation will waste their potential. This mindset creates inaction, as waiting for certainty becomes a substitute for movement. In reality, few paths are truly linear, and success is often the result of experimentation, iteration, and adaptation. Understanding that multiple routes can lead to meaningful outcomes frees students to take action. Momentum, not perfect foresight, is what drives growth.

The illusion of a perfect path is reinforced by external expectations. Teachers, parents, and peers may subtly communicate that only certain achievements are valuable. Students internalize these messages, narrowing their perception of opportunity. Comparing themselves to others' trajectories amplifies stress and indecision. By defining success on their own terms, students regain autonomy and clarity. Exploring multiple avenues allows for discovery and alignment with personal strengths and interests.

Fear of making the "wrong" choice often leads to procrastination. Students may endlessly research, plan, or seek advice before committing to any course of action. Each delay reinforces the idea that a perfect path exists and must be found before progress is made. In truth, most learning occurs through trial, error, and iterative action. Embracing flexibility and adaptability allows students to pivot when necessary. Each step taken provides feedback and refines understanding of what works.

The illusion of a single path also stifles creativity and experimentation. Students may avoid exploring unfamiliar interests or unconventional approaches out of fear that they will deviate from the "ideal" route. Limiting oneself to a single perceived path reduces exposure to new ideas, experiences, and skills. Recognizing that multiple routes can lead to success encourages curiosity and innovation. Experimentation becomes a tool for growth rather than a source of anxiety. Smart students expand their possibilities by taking action rather than waiting for certainty.

Finally, breaking free from the belief in one perfect path requires a mindset shift. Students must embrace uncertainty, focus on incremental progress, and value movement over flawless planning. Each decision, even if imperfect, contributes to growth and learning. Success emerges from consistent effort, adaptability, and resilience. Recognizing that there are many ways to achieve meaningful outcomes transforms indecision into action. Momentum replaces paralysis, and smart students learn that progress, not perfection, defines achievement.

QUOTES:

"Your work is going to fill a large part of your life, and the only way to be truly satisfied is to do what you believe is great work."
— Steve Jobs

Chapter .11

When Learning Becomes a Substitute for Doing

Smart students often fall into the trap of endless preparation, mistaking learning for actual progress. They may read books, watch tutorials, or collect advice in hopes of gaining the perfect solution before acting. While acquiring knowledge is valuable, it can become a form of procrastination when it prevents real-world application. Tasks are delayed, opportunities are missed, and momentum stalls while the mind remains in a cycle of preparation. Learning without execution creates the illusion of advancement, but results remain stagnant. Recognizing that action amplifies understanding is the first step to bridging the gap between knowledge and accomplishment.

One reason learning can replace doing is the desire for certainty. Students may feel that if they learn enough, they will avoid mistakes and make flawless decisions. This mindset creates paralysis because perfect preparation rarely exists. Real-world challenges are unpredictable, and competence emerges through iterative practice rather than theoretical mastery. Acting in the presence of uncertainty allows feedback to guide growth. Students who take measured steps while learning incrementally develop both confidence and capability. Experience becomes the ultimate teacher, complementing theoretical knowledge.

Overreliance on learning can also stifle experimentation and creativity. Students may focus on mastering established methods or frameworks instead of trying new approaches. They may avoid action that could expose gaps in knowledge, fearing mistakes or failure. This creates a narrow scope of problem-solving and limits the development of adaptive skills. By combining learning with active experimentation, students discover solutions and insights that theory alone cannot provide. Action transforms knowledge into applied intelligence.

The habit of substituting learning for doing can also undermine motivation. Students may feel productive while consuming information, yet achieve little tangible progress. This "illusion of productivity" reduces urgency to act and diminishes the satisfaction of accomplishment. Setting small, actionable goals ensures that knowledge is consistently applied and tested. Each step completed reinforces competence, builds confidence, and encourages continued effort. Momentum is generated not by what is learned alone, but by what is practiced and applied.

Endless learning can create a false sense of readiness. Smart students often convince themselves that acquiring more knowledge is equivalent to being fully prepared for action. They may delay projects, decisions, or experiments, believing they need perfect understanding before starting. In reality, preparation alone cannot replace the lessons learned through experience. Acting allows students to test assumptions, identify gaps, and refine their skills in context. Recognizing that readiness comes from doing empowers students to take meaningful steps forward.

Another consequence is reduced efficiency. Spending excessive time absorbing information leaves little time for practical application. Tasks that could be completed effectively remain in perpetual planning, slowing momentum. Students may feel busy and productive while producing minimal tangible results. Balancing learning with action ensures that effort contributes directly to progress. Applying knowledge in manageable steps allows students to integrate theory and practice simultaneously.

The habit of over learning can also diminish motivation. Students may feel that consuming information is progress, reducing the urgency to act. This "illusion of productivity" can lead to stagnation and frustration. Prioritizing action transforms motivation into sustained engagement, as tangible outcomes reinforce effort. Each small success demonstrates that learning is powerful when paired with doing. Momentum is fueled by applied knowledge, not by information alone.

Over learning can also reinforce fear of mistakes. Students may avoid trying new approaches because they feel unprepared or worry that errors will reveal inadequacy. This prevents valuable experience and slows skill development. Embracing mistakes as feedback fosters growth and resilience. Each attempt, successful or flawed, provides insight that deepens competence and understanding. Action becomes the bridge between theoretical knowledge and real-world achievement.

Over learning can also create mental fatigue. Constantly acquiring information without applying it consumes cognitive and emotional energy. Students may feel exhausted while producing little tangible progress. This drain reduces focus, motivation, and the capacity to act when opportunities arise. Balancing learning with practical application preserves energy and ensures effort produces meaningful results. Applying knowledge in manageable steps allows students to build competence without burning out.

Another consequence is the reinforcement of perfectionism. Students may feel they need to know everything before attempting a task, fearing mistakes or exposure of gaps in understanding. This mindset delays action and limits learning opportunities. Recognizing that mistakes are part of growth encourages experimentation and reduces anxiety. Each attempt provides valuable feedback and strengthens confidence. Knowledge becomes a tool for progress rather than a source of fear.

The habit of substituting learning for doing can also narrow perspective. Students may focus exclusively on acquiring knowledge in familiar areas, avoiding challenges that stretch their abilities. This creates a comfort zone that limits skill development and creativity. Combining learning with action encourages exploration and adaptive thinking. Engaging with new challenges allows students to apply knowledge in diverse contexts, deepening understanding and fostering innovation.

Over learning can also weaken the connection between effort and outcomes. Students may feel that time spent learning is equivalent to success, even when no tangible progress occurs. This can reduce accountability and the sense of achievement. Setting specific, action-oriented goals links effort to measurable results. Each completed task provides proof that knowledge is being converted into competence. Action reinforces learning and demonstrates that potential is being actively realized.

Over learning can also limit adaptability. Students who focus exclusively on mastering theory may struggle when real-world situations deviate from what they have studied. Rigid reliance on knowledge can prevent creative problem-solving and quick adjustments. Action exposes gaps, encourages flexibility, and strengthens the ability to respond to unexpected challenges. Integrating doing with learning allows students to adapt their understanding to practical situations. Momentum grows as knowledge is tested, refined, and applied in varied contexts.

Another effect is delayed feedback. When students remain in the learning phase, they postpone opportunities to receive real-world input. Without feedback, it's difficult to know which strategies are effective or which areas need improvement. Reflection on action provides insight that theoretical study alone cannot offer. Regular application ensures that learning is validated and adjusted as necessary. Students gain clarity about what works and what requires refinement. Action, coupled with reflection, creates a cycle of continuous improvement.

The habit of substituting learning for doing can also reduce confidence. Students may feel capable in theory but hesitant to act in practice, fearing mistakes will reveal inadequacy. Confidence develops when knowledge is tested through execution and results are observed. Small, deliberate actions provide proof of competence and build self-trust. Over time, repeated cycles of application reinforce both skill and confidence. Momentum increases as students experience firsthand that they are capable of producing meaningful outcomes.

Over learning may also reinforce indecision. Students may believe that further study is required to make the "right" choice, postponing action indefinitely. This delays progress and can create a sense of being stuck. Recognizing that action provides information and feedback allows decisions to be made more confidently. Movement produces clarity, whereas waiting for perfect knowledge often produces only anxiety. Experimentation and reflection turn uncertainty into informed, iterative learning. Action becomes a mechanism for both growth and discovery.

Perpetually substituting learning for doing can also create a false sense of security. Students may feel they are making progress simply by accumulating knowledge, even when their skills remain untested. This illusion can reinforce hesitation and delay meaningful application. True growth emerges only when knowledge is applied and refined through real-world practice. Acting provides feedback that theoretical learning cannot, bridging the gap between potential and achievement. Students who integrate doing with learning gain both competence and confidence.

Another consequence is stagnation in skill development. Over learning without execution prevents the transition from understanding to mastery. Students may know concepts deeply but lack the practical experience to implement them effectively. This gap slows momentum and creates frustration, as intelligence does not automatically translate into results. Intentional application of knowledge ensures that learning produces tangible growth. Each completed task reinforces skills and builds a foundation for future challenges.

Over learning can also increase the fear of failure. When students believe they must know everything before acting, any uncertainty feels risky. This mindset leads to hesitation and avoidance, limiting opportunities for experimentation and improvement. Embracing action as a learning tool reduces anxiety and strengthens resilience. Each step taken, even if imperfect, provides evidence of competence and contributes to ongoing development. Action becomes a practice ground where knowledge is tested, refined, and validated.

The habit of substituting learning for doing can also weaken time management. Excessive preparation consumes energy and delays tangible progress. Tasks that could be completed efficiently remain unfinished, creating stress and a sense of being "stuck." By allocating time to both learning and application, students ensure that effort translates into results. Small, deliberate actions bridge the gap between theory and practice. Each step forward generates momentum, confidence, and measurable achievement.

Finally, overcoming the trap of over learning requires a disciplined balance between study and action. Students must intentionally apply knowledge, reflect on outcomes, and iterate strategies to improve. Knowledge without application remains potential, while action without reflection limits insight. By combining learning with doing, smart students transform theory into skill, momentum, and meaningful results. The integration of action and learning turns intelligence into tangible achievement and unlocks the full potential of their abilities.

CHAPTER .12

THE COMFORT TRAP OF FAMILIAR STRENGTHS

Smart students often rely heavily on their natural talents, creating a comfort zone that limits growth. Skills that come easily are repeated, while more challenging areas are avoided. Over time, this reliance on familiar strengths can prevent the development of new competencies. Comfort feels safe, but it slows progress and reduces adaptability. True growth requires stepping beyond ease into the unfamiliar. Recognizing the trap is the first step toward expanding skills and embracing challenges.

The comfort trap is reinforced by early success. Students who excel in certain areas may receive praise and reinforcement, which encourages them to stick to what they know. This external validation strengthens the attachment to familiar abilities and discourages risk-taking. As a result, opportunities for broader development are often overlooked. By consciously seeking challenges outside their comfort zone, students can diversify skills and increase resilience. Growth is cultivated when talent is paired with deliberate effort in unfamiliar contexts.

Another consequence is stagnation in personal and professional growth. Relying solely on strengths limits exposure to new ideas, methods, and experiences. Students may become highly skilled in one area but underdeveloped in others, creating gaps that impede long-term success. Exploring new skills and perspectives expands capability and confidence. Deliberate practice in areas of weakness transforms comfort into competence. Students who embrace discomfort build versatility, making intelligence more adaptable and actionable.

The comfort trap can also reinforce fear of failure. Familiar skills feel safe because success is predictable, while trying new things carries risk. Students may avoid unfamiliar challenges to protect their reputation or self-image. This avoidance limits experience and slows momentum. Learning to tolerate discomfort and occasional failure is essential for growth. Each challenge outside the comfort zone strengthens skill, confidence, and resilience.

Relying on familiar strengths can also create a false sense of mastery. Students may believe they are fully competent because tasks in their comfort zone feel effortless. This overconfidence can prevent recognition of gaps and limit further development. True mastery requires pushing boundaries and confronting areas of weakness. Growth occurs when students intentionally engage with skills that challenge their current abilities. Expanding beyond comfort ensures that intelligence is applied broadly, not just in familiar contexts.

Another consequence is decreased creativity. When students stick to what they know, they often approach problems in predictable ways. Familiarity can breed mental rigidity, reducing the ability to innovate or generate new ideas. Exploring unfamiliar domains encourages experimentation and flexible thinking. By combining comfort with challenge, students discover new perspectives and strategies. Action outside the comfort zone fosters creativity, adaptability, and deeper problem-solving skills.

The comfort trap can also affect long-term career and personal development. Students may excel early in specific areas but lack the versatility needed to navigate complex environments. Overreliance on natural talent can limit exposure to emerging skills or opportunities. Deliberate practice in diverse areas prepares students for varied challenges and increases resilience. Embracing discomfort and uncertainty equips students to pivot effectively when circumstances change. Expanding capabilities ensures that intelligence remains a dynamic, useful tool.

Fear of failure is often magnified within the comfort trap. Students may hesitate to attempt new skills because mistakes feel riskier outside familiar territory. This avoidance reinforces the habit of sticking to strengths and slows overall growth. Learning to tolerate mistakes and view them as feedback transforms fear into motivation. Each attempt outside the comfort zone builds competence and confidence incrementally. Action, not avoidance, becomes the key to unlocking broader potential.

Overreliance on familiar strengths can also create complacency. Students may become satisfied with their natural abilities, believing that effort in challenging areas is unnecessary. This mindset slows growth and limits the development of new competencies. True progress requires consistent effort, especially in domains that feel difficult or unfamiliar. By intentionally confronting weaknesses, students expand their skill set and enhance adaptability. Comfort may feel safe, but purposeful discomfort is where meaningful learning occurs.

Another effect is limited problem-solving ability. When students rely solely on what comes easily, they may struggle to tackle complex or novel challenges. Familiar approaches may not apply in new contexts, leaving students unprepared for unexpected situations. Developing skills outside the comfort zone increases versatility and resilience. Experimenting with unfamiliar tasks encourages creative thinking and flexible strategies. Each experience beyond comfort enhances competence and prepares students for diverse challenges.

The comfort trap can also hinder personal growth. Students may avoid activities that push boundaries, missing opportunities for self-discovery and development. Skills that feel effortless do not stretch mental, emotional, or social capacities. Engaging in challenging experiences promotes confidence, resilience, and broader perspective. Action outside comfort zones fosters a mindset of continuous learning. Smart students who embrace challenges grow not only in skill but in adaptability and self-awareness.

Fear of judgment often reinforces the comfort trap. Students may worry that venturing into unfamiliar areas will expose gaps in ability or invite criticism. This concern reinforces reliance on safe, well-practiced skills. Learning to tolerate risk and potential failure encourages experimentation and learning. Each step into discomfort strengthens competence and self-trust. Over time, students learn that growth comes from action, not avoidance, and that mistakes are part of the process.

The comfort trap can also create a habit of avoidance. Students may instinctively gravitate toward tasks that feel easy, postponing or avoiding those that challenge them. Over time, this pattern reinforces skill gaps and narrows their range of competence. Growth requires intentional engagement with tasks that feel uncomfortable or unfamiliar. Each small challenge faced builds resilience and expands capability. By confronting avoidance, students transform comfort into opportunity for development.

Another consequence is a reduced ability to handle uncertainty. When students rely on familiar skills, unexpected situations can feel overwhelming or unmanageable. Familiarity breeds predictability, but life rarely offers guaranteed outcomes. Developing proficiency in new areas strengthens adaptability and decision-making under uncertainty. Experimentation and exposure to diverse experiences enhance confidence in navigating complex challenges. Action outside comfort zones prepares students for real-world unpredictability.

The comfort trap can also limit long-term ambition. Students may excel in their current domains but fail to explore broader possibilities. Staying within familiar strengths can create a plateau, where early success masks untapped potential. Deliberately pushing beyond comfort encourages continuous growth and exploration. Students expand their skill sets, develop new interests, and increase opportunities for achievement. Stepping beyond ease ensures intelligence is applied dynamically rather than statically.

Fear of imperfection often reinforces the comfort trap. Students may worry that attempting difficult tasks will reveal weaknesses or result in failure. This fear discourages experimentation and limits exposure to meaningful learning experiences. Embracing mistakes as part of the growth process allows students to act despite discomfort. Each challenge met builds competence, confidence, and resilience. Momentum grows when discomfort is approached intentionally rather than avoided.

Overreliance on familiar strengths can also create a misleading sense of competence. Students may feel confident in areas where they excel, overlooking gaps in knowledge or skill. This can lead to complacency and missed opportunities for growth. True competence comes from facing challenges outside of one's comfort zone. By tackling unfamiliar tasks, students develop versatility, adaptability, and resilience. Each successful step beyond comfort reinforces capability and expands the boundaries of what they can achieve.

Another effect is reduced learning potential. Focusing only on familiar skills limits exposure to new concepts and methods. Students may miss opportunities to discover interests or talents they did not realize they possessed. Engaging with new challenges fosters curiosity, experimentation, and creative problem-solving. Learning through discomfort strengthens both skill and confidence. Smart students who diversify their experiences maximize their potential and develop a broader toolkit for success.

The comfort trap can also hinder long-term achievement. Early reliance on natural strengths may bring immediate success, but over time it can prevent skill diversification needed for complex, real-world challenges. Developing competence in unfamiliar areas prepares students for diverse environments and unexpected opportunities. By intentionally confronting difficult tasks, students ensure sustainable growth. Adaptability becomes a core strength, complementing natural talent and enhancing long-term success. Each step beyond comfort contributes to cumulative capability.

Fear of failure is often magnified within the comfort trap. Students may avoid tasks that challenge them, fearing mistakes will expose weakness or inadequacy. This avoidance reinforces reliance on familiar skills and slows overall growth. Learning to tolerate mistakes and treat them as feedback fosters resilience and progress. Each experience outside comfort zones strengthens both skill and self-trust. Action, reflection, and iteration are essential for transforming potential into real-world competence.

Finally, escaping the comfort trap requires deliberate, consistent action. Students must identify areas of reliance on ease and purposefully practice skills outside their comfort zone. Incremental steps build adaptability, confidence, and competence over time. Momentum grows as challenges are met, mistakes are learned from, and progress is celebrated. Comfort transforms from a limitation into a launching point for growth. Smart students who embrace discomfort turn latent intelligence into dynamic, actionable achievement across all areas of life.

Chapter .13

Why Smart Students Avoid Beginner Status

Smart students often resist being beginners because they associate inexperience with failure or inadequacy. They fear looking incompetent in front of peers, teachers, or themselves. This mindset can prevent them from starting new activities, learning unfamiliar skills, or exploring different fields. Avoiding the beginner stage may feel protective, but it limits growth and opportunity. True learning requires humility, curiosity, and the willingness to start small. Recognizing that every expert was once a beginner is essential for breaking this mental barrier.

One reason students avoid beginner status is ego preservation. Being inexperienced can challenge self-perceptions of intelligence and competence. Students may feel vulnerable when starting something new because they anticipate mistakes or slow progress. This fear often leads to procrastination, over-planning, or avoidance of unfamiliar tasks. Embracing the beginner phase allows students to cultivate resilience and learn through experience. Accepting imperfection as part of growth transforms fear into motivation.

Another consequence is missed opportunities for skill development. Avoiding beginner status keeps students confined to familiar areas of expertise. They may excel in existing strengths but fail to develop competencies in new domains. Growth comes from stepping into unfamiliar territory and practicing skills from the ground up. Each small step taken as a beginner builds experience, confidence, and mastery over time. Action, even at a novice level, accelerates long-term learning and achievement.

Fear of judgment also reinforces avoidance. Students may worry that peers or mentors will perceive mistakes as incompetence. This concern creates hesitation and prevents engagement with challenges that could foster growth. By separating self-worth from immediate performance, students gain the freedom to experiment and learn. Mistakes become feedback rather than evidence of inadequacy. Starting as a beginner becomes an opportunity to practice, iterate, and improve without fear of judgment.

The avoidance of beginner status can also limit creativity and innovation. Students who refuse to start as novices may stick to familiar approaches, relying on existing knowledge and techniques. This prevents experimentation and the development of new solutions. True innovation requires exploration, trial, and error, which only occur when one is willing to start from scratch. Each small attempt in unfamiliar territory expands problem-solving abilities and encourages adaptive thinking. By embracing the beginner mindset, students cultivate creativity and discover new ways to apply their intelligence.

Another consequence is slower skill acquisition. Avoiding the early stages of learning delays exposure to feedback, practice, and refinement. Students may miss critical opportunities to build foundational skills that support advanced competence. Engaging with beginner-level tasks accelerates learning by providing a structured pathway for improvement. Small, consistent practice builds mastery over time, creating a solid base for complex problem-solving. Action at the novice stage generates both skill and confidence that cannot be achieved through observation alone.

Fear of imperfection often drives avoidance. Students may feel that being a beginner exposes inadequacy or threatens self-image. This fear discourages exploration and risk-taking, keeping them confined to familiar skills. Embracing mistakes as necessary steps in the learning process reduces anxiety and encourages action. Each error provides feedback that strengthens competence and resilience. Students learn that imperfection is not failure—it is the foundation of expertise.

The avoidance of beginner status can also create rigid thinking. Students may cling to established methods or knowledge because venturing into new areas feels uncomfortable. This limits flexibility and reduces adaptability when faced with complex or novel challenges. Practicing as a beginner encourages open-mindedness, curiosity, and problem-solving skills. Exposure to diverse experiences develops cognitive agility and increases confidence in tackling new tasks. Step-by-step engagement in unfamiliar areas expands both capability and perspective.

Avoiding beginner status can also foster unrealistic expectations. Students may expect immediate competence when trying something new, and any early struggle can feel like failure. This mindset creates frustration and may lead to abandoning new pursuits prematurely. Recognizing that initial difficulty is part of the learning curve allows students to persist. Growth requires patience, practice, and incremental improvement. Each small success as a beginner reinforces skill and encourages continued effort.

Another consequence is limited adaptability in unfamiliar situations. Students who avoid starting from scratch may struggle when confronted with challenges outside their expertise. Relying solely on familiar skills reduces problem-solving flexibility and slows adjustment to new contexts. Practicing as a beginner strengthens resilience and equips students to navigate uncertainty effectively. Each encounter with novelty builds confidence and expands capabilities. Adaptability becomes a natural byproduct of embracing early learning stages.

The avoidance of beginner status can also increase fear of taking risks. Students may hesitate to explore new interests, fearing that inexperience will reveal inadequacy. This reluctance narrows opportunities for growth, experimentation, and discovery. Accepting beginner status reduces fear and encourages proactive engagement. Risk-taking becomes a tool for learning rather than a source of anxiety. Students gain valuable experience that accelerates both skill development and personal growth.

Another effect is delayed accumulation of mastery. Avoiding the early stages of learning postpones the time it takes to achieve competence. Students may miss critical windows to develop foundational skills that support long-term achievement. Engaging with beginner-level challenges provides essential practice and feedback. Repetition and reflection accelerate skill acquisition and build confidence. Momentum grows as students embrace each step, transforming initial discomfort into cumulative expertise.

Avoiding beginner status can also create a dependency on external validation. Students may seek recognition for what they already know rather than embracing tasks where they are inexperienced. This focus on appearing competent prevents engagement with learning opportunities that feel risky or unfamiliar. Growth comes from prioritizing skill development over approval. Each effort made as a beginner strengthens autonomy and confidence. By valuing learning over validation, students cultivate resilience and momentum.

Another consequence is limited perspective. Relying solely on existing skills and knowledge can narrow understanding of broader contexts. Students may miss innovative solutions or alternative approaches because they avoid situations where they feel inexperienced. Engaging with new challenges expands thinking and encourages adaptive strategies. Exposure to diverse experiences develops both cognitive flexibility and problem-solving ability. Each step outside familiar territory contributes to growth and the capacity to handle complexity.

The avoidance of beginner status can also increase anxiety around performance. Students may feel pressure to excel immediately, which heightens fear of mistakes. This fear can prevent action and create a cycle of hesitation. By accepting that mistakes are a natural part of early learning, students reduce anxiety and enhance willingness to try. Incremental progress builds competence and demonstrates that initial imperfection is temporary. Action replaces fear, allowing intelligence to translate into tangible skill.

Another effect is slower development of confidence in unfamiliar domains. When students avoid starting as beginners, they delay the opportunity to experience mastery through practice. Confidence grows when effort is applied and results are observed over time. Deliberately engaging in new skills strengthens self-efficacy and reduces fear of failure. Each successful attempt reinforces momentum, skill, and belief in one's ability to learn. Momentum accumulates as students embrace the process of growth rather than the illusion of instant competence.

Avoiding beginner status can also stunt long-term ambition. Students may become overly focused on maintaining competence in familiar areas, neglecting the opportunity to explore broader horizons. This narrow focus limits skill diversification and slows growth over time. Engaging with new challenges, even when uncomfortable, opens doors to unexpected opportunities and achievements. Each step taken as a beginner expands capability and builds resilience. Momentum grows as students push past fear and embrace the process of learning from scratch.

Another consequence is a tendency to overanalyze decisions. Students may delay starting new projects because they fear appearing inexperienced or making mistakes. This overthinking prevents action and reduces the chance to gain practical experience. Starting as a beginner allows for experimentation, feedback, and iterative improvement. Action teaches lessons that theory alone cannot provide. Each attempt reinforces confidence and skill, gradually reducing fear of inexperience.

The avoidance of beginner status can also create frustration with growth itself. Students may feel stuck when progress in new areas is slower than in familiar domains. This impatience can lead to giving up prematurely, reinforcing reliance on comfort and existing strengths. Recognizing that skill acquisition requires time and effort encourages persistence. Small, consistent actions in unfamiliar areas produce steady improvement and long-term mastery. Momentum develops when students focus on progress rather than immediate perfection.

Fear of social judgment often compounds the problem. Students may worry about peers noticing their novice status, which discourages trying new things. This concern reinforces avoidance and prevents exposure to valuable learning experiences. By reframing mistakes as a natural part of growth, students reduce fear and increase willingness to experiment. Action and reflection replace hesitation and self-consciousness. Each step taken builds skill, resilience, and the confidence to explore new areas.

Finally, embracing beginner status requires intentional action and a growth-focused mindset. Students must accept initial discomfort, practice regularly, and view early mistakes as essential feedback. Each attempt outside comfort zones builds both competence and self-trust. Intelligence becomes actionable when paired with curiosity, humility, and persistence. By starting as beginners, smart students unlock latent potential, transform hesitation into momentum, and cultivate lifelong growth.

QUOTES:

"The journey of a thousand miles begins with a single step." — Lao Tzu

CHAPTER .14

THE MOTIVATION MYTH

Many smart students fall into the trap of waiting for motivation before taking action. They believe that they must feel inspired or energized to start a project or pursue a goal. This mindset often leads to procrastination and stagnation, as motivation rarely arrives on demand. In reality, action often precedes motivation, not the other way around. Momentum is created through consistent effort, not fleeting feelings of enthusiasm. Recognizing that motivation is the result of progress, not the cause, is essential for overcoming this barrier.

One reason the motivation myth persists is the allure of waiting for the "perfect moment." Students may delay starting tasks until conditions feel ideal or until they feel fully prepared. This often results in missed opportunities and extended periods of inactivity. Taking small, deliberate steps generates clarity and builds momentum, creating the conditions that spark genuine motivation. Action triggers energy and focus, while waiting often amplifies anxiety. By prioritizing movement over feelings, students reclaim control over their progress.

Another consequence is dependency on external factors. Students may rely on praise, deadlines, or peer pressure to fuel action, believing they cannot act without external motivation. This dependence limits autonomy and reduces the ability to sustain effort independently. Internal momentum is built through habit, structure, and deliberate action rather than fleeting inspiration. Each step completed reinforces competence and builds intrinsic motivation. Over time, students learn that action itself creates drive, not the other way around.

The motivation myth can also create cycles of frustration. Students wait to feel ready or energized, then criticize themselves for inactivity, further reducing motivation. This self-judgment reinforces hesitation and erodes confidence. By shifting focus from feeling motivated to taking deliberate action, students break this negative cycle. Small wins accumulate, generating both energy and confidence for subsequent steps. Momentum becomes self-reinforcing, replacing the need to wait for external inspiration.

Relying on motivation as a prerequisite for action can also undermine discipline. Students may wait for feelings of excitement or inspiration, thinking that effort without enthusiasm is wasted. This approach often leads to inconsistent work habits and incomplete projects. True discipline comes from showing up regularly, regardless of emotional state. Consistent action reinforces skill, builds momentum, and gradually generates motivation naturally. By focusing on routine and effort, students convert potential energy into measurable progress.

Another consequence is the overemphasis on inspiration over strategy. Students may spend more time searching for motivation than designing practical steps to achieve their goals. This misallocation of effort delays results and fosters frustration. Planning and structured action provide clarity and direction, reducing reliance on fluctuating feelings. Each completed step produces feedback and encourages continued effort. Action-focused strategies transform intelligence into tangible achievements, independent of transient emotions.

The motivation myth can also increase self-criticism. Students may feel guilty or inadequate when motivation is low, interpreting it as a personal failing. This mindset reinforces avoidance and reduces confidence in their ability to act. By separating worth from feelings, students can take action without waiting for inspiration. Momentum is built through movement, not emotional readiness.

Small, consistent actions accumulate into significant progress over time.

Dependence on motivation also limits adaptability. Students who wait to feel ready may miss opportunities that require immediate engagement. Life rarely aligns perfectly with emotional states, and waiting can result in stagnation. Embracing action as the primary driver ensures that progress continues despite uncertainty or discomfort. Each step taken fosters learning, competence, and confidence. Action creates conditions for motivation, rather than the other way around.

Waiting for motivation can also limit creativity and problem-solving. Students may postpone tackling challenges that require effort, assuming inspiration will guide them. This delay reduces opportunities to experiment, iterate, and discover new solutions. Action, even when not accompanied by motivation, provides insight and generates ideas that theory alone cannot produce. Each small step encourages adaptive thinking and reveals possibilities previously unnoticed. By acting first, students stimulate both creativity and competence simultaneously.

Another consequence is over-reliance on emotional cues. Students who believe they need to "feel motivated" may allow moods to dictate productivity. Good ideas, skills, and progress are often postponed until emotions align perfectly with tasks. This dependency can create inconsistency and undermine long-term achievement. Developing self-directed routines and focusing on deliberate effort bypasses the need for fluctuating emotional states. Each completed step reinforces confidence and builds momentum independent of mood.

The motivation myth can also contribute to procrastination. Waiting for inspiration is often a sophisticated form of avoidance, masking fear of failure or uncertainty. Students may tell themselves they cannot start until they feel ready, but readiness rarely comes spontaneously. Acting despite discomfort or hesitation is what generates clarity and progress. Small, repeated actions reduce anxiety and increase both competence and motivation over time. Momentum grows as students take initiative rather than waiting passively.

Dependence on motivation can also create unrealistic expectations for progress. Students may expect dramatic bursts of inspiration to produce immediate results. When this does not happen, they feel discouraged or disillusioned. Action-driven strategies create steady, incremental progress, which accumulates into meaningful achievement. By focusing on consistent effort rather than waiting for ideal emotions, students build lasting competence. Small wins reinforce confidence, creating the motivation they initially sought.

Waiting for motivation can also reduce resilience in the face of challenges. Students who rely on feeling inspired may give up when obstacles arise, believing they aren't "motivated enough" to continue. This mindset prevents the development of persistence, grit, and problem-solving skills. Action-focused strategies, however, train students to keep moving even when enthusiasm wanes. Each small success builds confidence and reinforces the ability to persist. Momentum becomes self-sustaining as effort, not emotion, drives progress.

Another consequence is an overemphasis on passion as a prerequisite for action. Students may avoid tasks or projects they perceive as unexciting, even when they are important for growth. Waiting for passion can lead to neglecting essential responsibilities and delaying skill development. Developing discipline and structure ensures consistent action regardless of emotional engagement. Each completed task reinforces competence and builds confidence. Action transforms intelligence into tangible outcomes even when passion is absent.

The motivation myth can also create a fragile sense of achievement. Students may feel that accomplishments are only valid if fueled by inspiration, dismissing progress made through effort and persistence. This mindset undervalues consistent work and delays momentum. Recognizing that achievement comes from deliberate action, not emotional peaks, allows students to appreciate their own efforts. Incremental progress produces meaningful results, reinforces capability, and gradually generates intrinsic motivation. Action becomes both the driver and measure of success.

Dependence on motivation may also amplify fear of failure. Students may avoid tasks until they feel "ready" to succeed perfectly, equating low motivation with inadequacy. This mindset reduces experimentation and limits learning opportunities. By acting despite discomfort or incomplete confidence, students learn that failure is a tool for improvement rather than a verdict on ability. Each attempt, regardless of outcome, contributes to skill, knowledge, and resilience. Momentum builds when movement replaces hesitation.

Waiting for motivation can also create unnecessary stress and self-judgment. Students may berate themselves for feeling "lazy" or unmotivated, believing that their lack of inspiration reflects personal failure. This internal pressure often intensifies procrastination rather than resolving it. Recognizing that motivation is a byproduct of action, not a prerequisite, reduces self-criticism. Each deliberate step taken builds confidence, competence, and a sense of control. Momentum grows as students replace judgment with consistent, purpose-driven action.

Another consequence is the illusion of readiness. Students may convince themselves that they are not prepared until motivation strikes, postponing valuable experience. In reality, readiness comes through action and iteration, not waiting for ideal feelings. Starting despite uncertainty produces feedback that accelerates learning and builds mastery. Each attempt provides insight, strengthens capability, and fuels further movement. By acting first, students cultivate momentum that generates both motivation and clarity.

The motivation myth can also diminish long-term achievement. Students may achieve less because they wait for inspiration rather than creating progress through structured effort. Relying on motivation creates inconsistency, which compounds over time and limits cumulative growth. Prioritizing action over emotion ensures steady, sustainable results. Each small success builds confidence, reinforces competence, and produces the intrinsic drive needed for larger goals. Momentum accumulates as effort becomes habitual rather than conditional on feeling inspired.

Dependence on motivation may also prevent students from experiencing the full value of discipline. Waiting for feelings undermines the development of habits, routines, and work ethic that are essential for consistent achievement. By acting regardless of motivation, students strengthen self-regulation and internal accountability. Each completed action reinforces autonomy and reduces reliance on external or internal emotional cues. Smart students learn that progress is driven by effort, not emotion, and that discipline produces both momentum and motivation.

Finally, overcoming the motivation myth requires a shift from emotion-driven to action-driven practice. Students must focus on consistent effort, structured routines, and incremental goals to sustain progress. Motivation follows action, building confidence and reinforcing competence. By taking initiative first and allowing inspiration to emerge naturally, students convert intelligence into tangible outcomes. Momentum, achievement, and growth become self-sustaining as action replaces waiting. Smart students unlock their potential when they move first and let motivation catch up.

QUOTES:

"You don't have to be great to start, but you have to start to be great." — Zig Ziglar

CHAPTER .15

THE APPROVAL LOOP

Smart students often fall into the approval loop, constantly seeking validation from teachers, peers, or authority figures. They may measure their self-worth by grades, praise, or recognition, rather than their own standards of growth. This dependence on external approval can limit creativity, risk-taking, and authentic decision-making. When action is driven by the desire to please others, momentum slows, and personal goals are neglected. True autonomy comes from aligning effort with personal values rather than external expectations. Recognizing the approval loop is the first step toward breaking free and reclaiming agency over one's learning journey.

One reason students enter the approval loop is the conditioning of traditional education systems. Schools often reward compliance, speed, and correctness, reinforcing the habit of seeking approval. Students internalize the idea that success is defined by others, rather than self-directed progress. Over time, this mindset creates anxiety, self-doubt, and hesitation to pursue unfamiliar challenges. By consciously shifting focus from external validation to intrinsic standards, students regain control over learning and achievement. Momentum and confidence flourish when effort is guided by personal goals rather than outside judgment.

Another consequence is stifled innovation. Seeking constant approval discourages experimentation and risk-taking, as students fear negative feedback or disapproval. This leads to predictable outcomes and limited growth, even for highly intelligent students.

Authentic creativity and problem-solving require the freedom to make mistakes and explore new approaches. By prioritizing personal standards over external validation, students develop resilience, curiosity, and adaptive thinking. Action becomes purposeful, and results become evidence of competence rather than approval.

The approval loop can also increase anxiety and self-criticism. Students may worry about how their work will be perceived, leading to overthinking or procrastination. They may avoid challenges where success is uncertain or judgment is likely. Breaking the loop requires detachment from external opinions and a focus on personal growth. Reflection, goal-setting, and self-assessment become tools to measure progress instead of others' recognition. Momentum grows as students act with intention, free from the constraints of seeking approval.

Constantly seeking approval can also reduce focus on long-term goals. When students prioritize pleasing others, they may take on tasks or projects that do not align with their personal vision. This diversion of effort slows progress toward meaningful achievement. By shifting attention to intrinsic objectives, students ensure that energy and time are invested in pursuits that truly matter. Each action aligned with personal goals builds momentum, competence, and satisfaction. Freedom from external validation allows students to move confidently toward long-term growth.

Another consequence is the reinforcement of self-doubt. Students who depend on approval often question their abilities when praise is absent. They may feel inadequate or incapable, even when objectively competent. By developing internal standards of success, students cultivate self-trust and resilience. Confidence emerges from evidence of action and results rather than the opinions of others. Each step taken independently strengthens competence and reduces reliance on external validation.

The approval loop can also limit risk-taking and experimentation. Fear of criticism or disapproval encourages students to stick to safe, familiar tasks. This stifles innovation and slows skill development, even among highly capable learners. Embracing action aligned with personal standards encourages exploration and learning from mistakes. Each challenge attempted provides feedback, builds skill, and increases resilience. Momentum grows when students take bold steps independent of others' expectations.

Dependence on approval may also affect decision-making. Students may defer choices or alter plans to meet perceived expectations, rather than following their judgment. This compromises autonomy and can lead to frustration or stagnation. By prioritizing personal criteria for action, students regain control over decisions. Each decision made in alignment with internal goals reinforces confidence and forward momentum. Autonomy becomes a powerful driver of sustained achievement.

Seeking approval can also create a reliance on comparison. Students may constantly measure themselves against peers or benchmarks, rather than focusing on their own progress. This comparison can breed envy, self-doubt, and distraction from meaningful learning. By shifting focus inward, students can track growth relative to themselves instead of others. Each step of personal progress reinforces competence and confidence. Momentum builds when effort is measured by improvement, not by external validation.

Another consequence is reduced intrinsic motivation. When actions are guided by the desire for praise or recognition, students may feel less driven by curiosity or personal interest. This dependence on external validation undermines engagement and long-term commitment. Cultivating internal motivation through personal goals and self-directed challenges restores agency. Each accomplishment, no matter how small, reinforces the value of effort and fuels further progress. Motivation becomes sustainable when it emerges from within rather than being dictated by others.

The approval loop can also limit resilience in the face of criticism. Students accustomed to constant validation may take feedback personally, interpreting it as a threat to their self-worth. This can increase anxiety and reduce willingness to try new or challenging tasks. By reframing feedback as information rather than judgment, students can respond constructively. Each encounter with critique becomes an opportunity to learn and strengthen skill. Momentum and confidence increase as students act despite potential disapproval.

Dependence on approval may also create hesitation to innovate. Students may avoid unconventional approaches if they fear negative evaluation. This limits creativity and the ability to develop unique solutions to problems. By prioritizing personal standards over external opinions, students can take thoughtful risks. Each innovative step, even if imperfect, builds competence and confidence. Momentum grows when students act authentically and embrace experimentation.

Dependence on approval can also slow decision-making. Students may hesitate to choose a direction, fearing that others will judge or disagree with their choices. This indecision delays progress and reduces confidence in their own judgment. By learning to trust their internal criteria, students can make decisions efficiently and confidently. Each choice aligned with personal goals reinforces autonomy and competence. Momentum builds as action replaces hesitation, allowing progress to accelerate despite external opinions.

Another consequence is the reinforcement of perfectionism. Students seeking approval may feel pressure to produce flawless work to earn recognition. This fear of imperfection can prevent experimentation, limit learning, and increase stress. Embracing action for the sake of growth rather than praise encourages risk-taking and resilience. Each attempt, regardless of outcome, strengthens skills and builds confidence. Momentum develops as students prioritize progress over external validation or unattainable perfection.

The approval loop can also stifle creativity and innovation. Students may avoid unconventional ideas, fearing disapproval or criticism. This limits exploration, experimentation, and the development of novel solutions. Prioritizing internal standards over external judgment empowers students to take risks and innovate. Each creative step expands capability, resilience, and confidence. Momentum grows when action is guided by purpose rather than the need for approval.

Dependence on approval can also reduce intrinsic satisfaction. Students may achieve milestones yet feel empty if external validation is lacking. This undermines engagement, motivation, and the sense of accomplishment. By focusing on personal growth and self-defined success, students experience fulfillment from effort and progress. Each achievement, no matter how small, reinforces self-worth and fuels continued action. Momentum strengthens when satisfaction comes from meaningful progress rather than external praise.

Relying on approval can also create emotional dependency. Students may tie their moods and self-esteem to the reactions of others, creating volatility in confidence. Praise temporarily boosts self-worth, while criticism can lead to discouragement or doubt. Developing internal validation through personal standards stabilizes confidence and emotional resilience. Each action taken independently reinforces competence and self-trust. Momentum grows as students learn to act consistently regardless of external approval.

Another consequence is the distraction from meaningful goals. Students may prioritize tasks that earn recognition over those that align with their personal vision or growth. This misalignment diverts time and energy from impactful work. By focusing on intrinsic objectives, students ensure that effort contributes directly to long-term progress. Each deliberate action toward personal goals builds skill, confidence, and momentum. Alignment between effort and purpose fosters sustainable growth and fulfillment.

The approval loop can also increase stress and anxiety. Constantly seeking validation can make students hyper-aware of judgment, creating pressure and mental fatigue. This stress reduces focus, creativity, and enjoyment in learning. By shifting attention from external approval to personal standards, students reduce unnecessary tension and improve performance. Each autonomous action strengthens confidence and resilience. Momentum emerges as effort becomes driven by purpose, not fear.

Dependence on approval may also limit personal expression. Students may avoid unconventional ideas or approaches, fearing disapproval. This reduces creativity, innovation, and authentic engagement. By prioritizing personal judgment and goals over external feedback, students gain freedom to experiment. Each step taken according to internal standards fosters skill, confidence, and originality. Momentum builds as authentic action replaces the need for constant validation.

Finally, breaking free from the approval loop requires conscious effort, reflection, and consistent practice. Students must set personal standards, act purposefully, and detach self-worth from external reactions. Feedback can inform but should not control action. Confidence, autonomy, and sustainable progress grow when effort aligns with internal values. Smart students unlock their potential by acting independently, embracing authentic growth, and escaping the cycle of seeking approval.

QUOTES:

"Don't let the noise of others' opinions drown out your own inner voice." — Steve Jobs

Chapter .16

Burnout Without Results

Many smart students experience burnout despite putting in long hours and immense effort. They may feel exhausted but see little progress, creating frustration and discouragement. This often happens when effort is misaligned with strategy or focused on tasks that don't produce meaningful outcomes. Working harder without direction is not the same as working effectively. Burnout without results erodes motivation, confidence, and well-being. Recognizing the signs early is crucial to redirect effort toward actions that generate real progress.

One reason burnout occurs is overcommitment. Students take on too many responsibilities, believing they must excel in all areas simultaneously. This can lead to mental, emotional, and physical exhaustion without substantial achievement. Prioritizing tasks and focusing on high-impact activities ensures energy is invested where it counts most. Each strategic step produces measurable results, reinforcing competence and confidence. Momentum grows when effort is aligned with meaningful goals rather than sheer volume.

Another cause is inefficient learning and working methods. Students may spend excessive time on preparation, perfectionism, or low-priority tasks. Hours of effort may be invested without producing tangible progress, contributing to exhaustion and frustration. Developing effective systems, routines, and processes maximizes output while conserving energy. Each completed task builds both skill and a sense of accomplishment. Smart effort, rather than raw effort, creates sustainable momentum and prevents burnout.

The pressure to maintain constant performance also fuels burnout. Students may feel they must consistently demonstrate intelligence and competence, leading to relentless self-imposed standards. This pressure amplifies stress and reduces the ability to recover between tasks. Learning to set realistic expectations and embrace incremental progress protects energy and focus. Small, deliberate actions produce results without overtaxing mental and emotional resources. Momentum is built by sustainable effort, not continuous overextension.

Overworking without results can also erode motivation over time. Students may feel that no matter how much effort they put in, their progress remains stagnant. This perception can create discouragement and a sense of futility. By focusing on actions that produce measurable outcomes, students restore motivation and purpose. Each completed task, no matter how small, reinforces competence and confidence. Momentum grows as effort translates into tangible achievements rather than wasted energy.

Another consequence is impaired focus and decision-making. Exhaustion from excessive effort reduces cognitive clarity and increases mistakes. Students may struggle to prioritize effectively or choose optimal strategies, further limiting results. Developing structured routines and time-management strategies ensures energy is directed toward high-impact activities. Each focused action produces visible progress, building both skill and confidence. Strategic effort prevents wasted energy and reduces the risk of burnout.

Burnout without results can also harm self-esteem. Students may equate exhaustion and lack of progress with personal inadequacy or failure. This internalization undermines confidence and can perpetuate cycles of stress and avoidance. By tracking progress objectively and celebrating incremental wins, students reinforce their sense of competence. Each measurable accomplishment strengthens self-trust and motivation. Momentum develops when action is tied to results rather than effort alone.

The pursuit of perfection often compounds burnout. Students may spend excessive time refining work beyond what is necessary, delaying outcomes and draining energy. This relentless pursuit of flawlessness prevents forward movement and contributes to mental fatigue. Emphasizing progress over perfection ensures that effort produces meaningful results. Each completed task, even imperfectly, builds skill, confidence, and momentum. Smart students learn that action and iteration outweigh endless refinement without output.

Burnout without results can also reduce creativity and problem-solving ability. When students are exhausted from unproductive effort, their minds have less energy to explore new ideas or approaches. Mental fatigue limits insight, innovation, and adaptive thinking, making it harder to achieve meaningful outcomes. By directing energy toward focused, high-impact work, students preserve mental capacity for creative problem-solving. Each completed task reinforces clarity and confidence in decision-making. Momentum grows when effort is purposeful rather than wasted on low-yield activity.

Another consequence is increased stress and anxiety. Students may feel pressure to perform continuously without seeing progress, creating tension and worry. This emotional burden can interfere with focus, learning, and well-being. Implementing structured routines, prioritization, and strategic breaks reduces stress while maintaining productivity. Each action that produces measurable results builds reassurance and reduces the mental strain of uncertainty. Momentum emerges as strategic work replaces reactive, exhausting effort.

Overextension can also impact physical health. Long hours of mental or academic work without adequate rest or recovery can lead to fatigue, sleep disruption, and decreased energy. Physical depletion further limits cognitive performance and learning potential. Incorporating regular rest, movement, and self-care ensures that energy levels remain sustainable. Each balanced approach supports continued effort and measurable outcomes. Momentum is maintained when both mind and body are cared for strategically.

The fear of missing out or falling behind can exacerbate burnout. Students may push themselves to stay ahead or outperform peers, even when their efforts do not produce tangible progress. This pressure contributes to exhaustion and diminished results. By focusing on high-value actions aligned with personal goals, students reduce unnecessary stress and increase meaningful output. Each purposeful effort builds competence and confidence while maintaining energy. Momentum accelerates when effort is aligned with results, not competition or comparison.

Burnout without results can also erode long-term engagement and passion. Students who repeatedly exert effort without seeing progress may begin to question their abilities or interest in the subject. This disengagement can lead to withdrawal, reduced effort, and missed opportunities for growth. By focusing on actions that produce tangible outcomes, students reinforce motivation and maintain enthusiasm. Each visible achievement, no matter how small, restores confidence and commitment. Momentum grows as effort begins to yield meaningful results rather than exhaustion alone.

Another consequence is a diminished sense of control. When effort doesn't translate into results, students may feel powerless, believing that success is dictated by luck, circumstance, or external factors. This perception undermines autonomy and confidence in decision-making. Establishing systems, priorities, and feedback loops allows students to regain control over outcomes. Each intentional, measurable step reinforces agency and accountability. Momentum is strengthened when students recognize the direct connection between strategic action and meaningful results.

Overworking without results can also create negative emotional cycles. Frustration, anxiety, and self-criticism often build when effort seems futile, feeding further exhaustion. These emotions can reduce focus, creativity, and resilience. By structuring effort toward high-impact tasks and tracking progress, students break the cycle of wasted energy. Each completed task, even incremental, reinforces confidence and reduces emotional strain. Momentum grows as strategic action replaces the reactive, draining patterns that fuel burnout.

The pursuit of perfection often compounds the problem. Students may spend excessive time polishing work, fearing mistakes or disapproval, which delays outcomes and drains energy. This perfectionism contributes to both mental fatigue and lack of measurable results. Prioritizing completion and progress over flawless execution ensures energy is used effectively. Each completed project, regardless of minor imperfections, builds skill, confidence, and momentum. Smart students learn that forward movement matters more than unattainable perfection.

Burnout without results can also distort students' perception of their own abilities. When long hours of effort fail to produce visible outcomes, students may begin to doubt their intelligence or capacity to succeed. This self-doubt can lead to hesitation, reduced effort, and avoidance of new challenges. By focusing on measurable progress rather than effort alone, students reinforce confidence and competence. Each completed step validates ability and encourages continued action. Momentum grows as effort aligns with tangible results, restoring trust in one's own skills.

Another consequence is the development of inefficient habits. Students may repeat the same unproductive behaviors, believing that persistence alone will eventually yield results. Without reflection and strategy, effort becomes busywork rather than meaningful action. Implementing deliberate routines, prioritization, and feedback loops ensures that effort translates into measurable outcomes. Each action taken strategically builds skill, understanding, and confidence. Momentum accelerates when smart effort replaces mindless persistence.

Burnout without results can also reduce emotional resilience. Continuous effort without visible progress can lead to frustration, irritability, and stress. These emotional states interfere with learning, focus, and creativity. By aligning effort with clear, achievable objectives, students reduce stress and reinforce a sense of accomplishment. Each successful step strengthens both skill and emotional endurance. Momentum grows as effort produces tangible outcomes and emotional balance is maintained.

The illusion of constant busyness often masks burnout. Students may feel productive because they are always working, yet little progress is actually made. This creates a false sense of accomplishment and prolongs frustration. Focusing on high-impact tasks and measurable results ensures that time and energy are well-spent. Each completed, meaningful task builds competence, confidence, and forward momentum. Smart students learn that action without results is empty, and strategic effort is what sustains growth.

Finally, overcoming burnout without results requires intentional reflection, strategic effort, and sustainable habits. Students must identify tasks that produce tangible outcomes, implement effective routines, and integrate rest to maintain energy and focus. Each deliberate action reinforces progress, competence, and confidence. Momentum is generated when effort is purposeful, measurable, and sustainable. By working strategically and aligning energy with meaningful results, smart students prevent exhaustion, accelerate growth, and turn intelligence into consistent achievement.

QUOTES:

"Don't confuse activity with productivity."

— Robin Sharma

Part III: Rebuilding How Smart Students Think

CHAPTER .17

REDEFINING WHAT PROGRESS ACTUALLY LOOKS LIKE

Many smart students struggle because they measure progress by perfection rather than movement. They may feel stuck when outcomes aren't flawless, overlooking the value of incremental improvement. True progress is about consistent action, learning from mistakes, and refining skills over time. By shifting focus from mastery to measurable steps, students create momentum and maintain motivation. Each small success contributes to cumulative growth, building both competence and confidence. Understanding progress as movement rather than perfection allows smart students to act with clarity and purpose.

One reason students misinterpret progress is the education system's emphasis on grades and performance. Early success often rewards speed, correctness, and memorization, creating the illusion that progress is linear. In reality, meaningful growth is nonlinear, with setbacks, experiments, and repeated practice as integral parts. Tracking small wins, reflecting on learning, and celebrating incremental achievements reframes how students view their development. Each step forward, even if imperfect, adds to long-term competence. Momentum builds when action and reflection replace comparison and unrealistic expectations.

Another consequence of mismeasuring progress is discouragement. Students may feel they aren't advancing if outcomes don't match their high standards, even when effort and learning are substantial. This can create stagnation, self-doubt, and reduced initiative. By redefining progress as consistent effort and improvement, students maintain motivation and resilience. Each small accomplishment validates competence and reinforces the habit of forward action. Momentum emerges as movement, not perfection, becomes the standard for success.

Focusing on movement over mastery also encourages experimentation and adaptability. Students become more willing to try new strategies, explore unfamiliar skills, and learn from mistakes. This approach fosters growth, resilience, and creativity. Each trial, regardless of immediate success, provides feedback that informs future decisions. Momentum is created when action generates knowledge, confidence, and adaptability simultaneously. Redefining progress ensures smart students continue to evolve even when outcomes aren't immediate or perfect.

Measuring progress by movement rather than perfection also reduces fear of failure. Students often hesitate to start or experiment because they equate mistakes with lack of ability. This fear can slow growth and prevent skill development. By valuing incremental action, students learn that errors are part of the process and provide important feedback. Each attempt, regardless of outcome, builds competence and confidence. Momentum grows as students focus on learning from experience instead of avoiding mistakes.

Another benefit is sustained motivation. When progress is defined as consistent steps rather than immediate mastery, students are less likely to feel discouraged. They recognize that each small effort contributes to long-term growth. Celebrating minor achievements reinforces intrinsic motivation and encourages continued action. Each completed step, no matter how small, accumulates into meaningful results. Momentum is maintained as focus shifts from ideal outcomes to tangible movement forward.

Redefining progress also enhances clarity in goal-setting. Students can break large objectives into smaller, actionable steps, making challenges less overwhelming. This approach allows for measurable outcomes and a sense of accomplishment along the way. Each milestone, even incremental, provides evidence of growth and capability. Momentum builds as students consistently move toward goals with achievable, concrete actions. Focusing on manageable progress transforms large ambitions into attainable results.

This mindset also encourages reflection and self-assessment. Students can evaluate what is working and what needs adjustment without feeling defeated by imperfections. By tracking incremental improvement, they can make strategic decisions about next steps. Each reflection session enhances understanding, efficiency, and skill acquisition. Momentum is reinforced when students learn to adapt based on evidence of progress rather than fear of failure. Growth becomes both measurable and sustainable.

Redefining progress also encourages long-term vision. Students who focus only on immediate outcomes may overlook the bigger picture of skill development and personal growth. By measuring movement over time, they can see how incremental actions accumulate into significant achievements. Each small step contributes to building expertise, confidence, and resilience. Momentum strengthens as students recognize that consistent effort compounds over time. This perspective allows them to stay motivated even when immediate results are not dramatic.

Another consequence of mismeasuring progress is the temptation to compare oneself to others. Students may feel inadequate when peers achieve visible success faster, even if their own efforts are consistent and meaningful. Focusing on personal movement rather than external benchmarks reduces the stress of comparison. Each step taken in alignment with one's own goals reinforces competence and autonomy. Momentum grows as students measure progress by their own trajectory instead of others' outcomes. Growth becomes sustainable and self-directed.

Valuing movement over mastery also cultivates adaptability. Students who embrace incremental progress are more willing to adjust strategies, learn from mistakes, and explore alternative approaches. This flexibility enhances problem-solving and innovation. Each attempt, even if imperfect, provides insight that informs the next step. Momentum builds as students learn to pivot and respond effectively rather than being paralyzed by fear of failure. Redefining progress creates a mindset that thrives on learning and iteration.

This approach also enhances confidence. Students see that consistent action, rather than instant perfection, produces tangible results. Each completed task validates ability, reduces self-doubt, and reinforces belief in one's capacity to grow. Momentum is reinforced as confidence encourages further action, creating a self-perpetuating cycle of progress. By recognizing that skill develops incrementally, smart students act boldly, experiment, and take initiative. Progress becomes evidence-based rather than aspirational or hypothetical.

Redefining progress also encourages persistence during setbacks. Students often feel stalled when immediate results don't match effort, leading to frustration or giving up. By valuing incremental movement, they can continue acting even when outcomes are delayed or imperfect. Each small effort reinforces learning, skill development, and resilience. Momentum is maintained as students recognize that consistent action leads to cumulative growth. Viewing progress as ongoing movement rather than immediate success helps smart students stay engaged and persistent.

Another benefit is improved focus on high-impact actions. When progress is measured by movement rather than perfection, students can prioritize tasks that generate tangible results. This prevents wasted effort on low-priority activities or endless refinement. Each action taken with purpose contributes to measurable achievement and skill development. Momentum builds as students consistently act on tasks that matter most. Focusing on meaningful movement ensures energy is directed where it produces real growth.

This mindset also fosters realistic expectations. Students learn that mastery takes time, practice, and repeated effort. They understand that each step, even small or imperfect, contributes to competence. This reduces frustration and self-criticism while encouraging continued action. Each completed task reinforces confidence and sustains momentum. Redefining progress helps students develop patience and a long-term perspective, making achievement more attainable.

Redefining progress can also strengthen self-efficacy. By tracking incremental wins and recognizing the impact of consistent effort, students see that their actions directly influence outcomes. This realization empowers them to take initiative and tackle challenges proactively. Each step taken validates their ability to grow and succeed. Momentum grows as confidence in one's capacity to make progress reinforces further action. Smart students internalize that consistent effort leads to mastery over time.

Redefining progress also promotes a healthier relationship with failure. Students who focus only on outcomes may see mistakes as evidence of inadequacy, rather than opportunities to learn. By valuing movement and incremental improvement, failures are reframed as feedback rather than judgment. Each misstep becomes a guide for adjustment and growth. Momentum grows as students continue taking action, learning from experience rather than being paralyzed by fear. Smart students develop resilience and confidence by understanding that progress is measured by effort and adaptation, not perfection.

Another benefit is enhanced long-term vision and planning. When students track small, consistent steps, they can see how short-term actions build toward larger goals. This clarity helps maintain motivation and ensures that effort aligns with meaningful objectives. Each small achievement confirms that consistent movement leads to cumulative success. Momentum is reinforced as students recognize the compounding effect of repeated action. By focusing on incremental progress, smart students bridge the gap between immediate effort and long-term mastery.

Focusing on movement also strengthens self-compassion. Students learn to appreciate effort, persistence, and learning as valuable in themselves, even if results are delayed. This reduces stress, comparison, and harsh self-judgment. Each step forward, regardless of perfection, reinforces confidence and a sense of competence. Momentum grows as students celebrate progress rather than punish themselves for imperfection. A movement-based perspective fosters resilience, satisfaction, and sustained engagement.

Redefining progress encourages continuous learning and experimentation. When the focus is on movement, students are more willing to try new approaches, explore unfamiliar skills, and embrace challenges. Each attempt provides feedback, strengthens skill, and builds knowledge. Momentum builds as students iterate, adapt, and grow from each experience. Smart students develop a mindset that thrives on learning and action, rather than waiting for perfect outcomes.

Finally, redefining progress requires intentional practice, reflection, and mindset shifts. Students must track small wins, focus on consistent effort, and reframe mistakes as learning opportunities. Each step forward strengthens skill, confidence, and momentum. Momentum is sustained when progress is measured by incremental action rather than flawless results. By redefining progress, smart students transform intelligence into continuous growth, maintain motivation, and achieve meaningful, lasting results.

QUOTES:

"Success is the sum of small efforts, repeated day in and day out."
— Robert Collier

CHAPTER .18

SEPARATING SELF-WORTH FROM PERFORMANCE

Many smart students tie their self-worth directly to achievements, grades, or accolades. This can create a fragile sense of identity, where success feels like validation and failure feels like personal inadequacy. When self-esteem is contingent on performance, stress and anxiety increase, making learning and growth more difficult. Separating self-worth from outcomes allows students to take risks, make mistakes, and explore without fear of judgment. Confidence becomes rooted in identity, curiosity, and effort rather than external validation. By detaching personal value from performance, smart students can act freely and cultivate sustainable momentum.

One reason this attachment forms is early reinforcement from school and society. Students are often praised for intelligence, speed, or results rather than effort, creativity, or resilience. Over time, they internalize the belief that their value depends on achievement. This mindset discourages experimentation, risk-taking, and embracing challenges. Learning to value effort, growth, and curiosity allows students to develop a more stable sense of self-worth. Each action taken without attaching identity to outcome strengthens confidence and autonomy.

Another consequence of tying self-worth to performance is fear of failure. Students may avoid difficult tasks, novel challenges, or opportunities that could produce imperfect results. This limits growth and reinforces a cycle of safe, familiar choices. By recognizing that mistakes are part of learning, students can act without self-judgment. Each step taken independent of outcome reinforces competence and reduces fear. Momentum builds as action becomes fueled by curiosity and growth rather than fear of judgment.

Separating self-worth from performance also improves resilience. Students who define themselves by results may experience intense discouragement when progress stalls or setbacks occur. This can lead to stress, burnout, and disengagement. By valuing identity beyond outcomes, students maintain confidence and continue taking action through challenges. Each small success, regardless of external recognition, reinforces growth and capability. Momentum emerges as effort is sustained, even when results fluctuate or challenges arise.

Focusing too heavily on outcomes can also distort decision-making. Students may choose tasks or goals based on what will look impressive rather than what aligns with their growth or interests. This can lead to misaligned effort, frustration, and slower skill development. By separating self-worth from performance, students can pursue challenges that truly matter to them. Each choice made for personal growth rather than validation reinforces autonomy and purpose. Momentum builds as action is guided by intrinsic goals rather than external approval.

Another consequence is the amplification of self-criticism. When self-esteem depends on achievement, mistakes and setbacks are interpreted as personal failures. This harsh self-judgment reduces confidence, motivation, and resilience. By detaching identity from performance, students can evaluate results objectively and respond with constructive adjustments. Each step taken becomes a learning opportunity rather than a measure of worth. Momentum is maintained as effort continues despite setbacks, reducing the paralyzing fear of failure.

The attachment of self-worth to performance also encourages overwork and perfectionism. Students may feel compelled to over-prepare or overachieve to "prove" their value, often at the expense of health and balance. This cycle leads to exhaustion and diminished results. By separating identity from outcomes, students can act strategically and sustainably, prioritizing learning and growth over proving themselves. Each deliberate effort contributes to competence and confidence rather than external validation. Momentum grows when action is focused on mastery and experience instead of approval.

Separating self-worth from performance also strengthens intrinsic motivation. Students begin to pursue tasks because they find them meaningful or interesting, rather than for praise or recognition. This shift increases engagement, creativity, and resilience, as effort is fueled by curiosity and personal goals. Each action reinforces self-directed growth and builds sustainable momentum. Momentum compounds as students act consistently, independent of external judgment or fluctuating results. Intelligence becomes actionable when paired with purpose and autonomy.

Detaching self-worth from performance also reduces anxiety around evaluation. Students who tie identity to results often obsess over grades, feedback, or comparisons. This stress can impair focus, creativity, and enjoyment in learning. By separating self-esteem from outcomes, students can approach tasks with curiosity and purpose instead of fear. Each action becomes an opportunity to learn rather than a test of value. Momentum grows as effort is fueled by interest and engagement rather than pressure to prove oneself.

Another benefit is improved risk-taking and experimentation. Students who define themselves by results may avoid unfamiliar or challenging tasks, fearing failure will diminish their worth. This limits growth and exploration. By valuing effort and learning over outcomes, students can take calculated risks and pursue new experiences. Each attempt builds skill, knowledge, and confidence, regardless of immediate success. Momentum builds as action is guided by curiosity and growth rather than fear of judgment.

Separating self-worth from performance also encourages reflection and adaptation. When identity is not tied to results, students can evaluate successes and failures objectively, learning from each experience. This promotes continuous improvement and resilience. Each reflection enhances understanding and informs strategic decisions for future actions. Momentum is reinforced as students act with awareness and adaptability, rather than reacting emotionally to outcomes. Growth becomes a deliberate, sustainable process.

This mindset also enhances emotional stability. Students who internalize self-worth from sources beyond performance experience fewer highs and lows tied to outcomes. They are less likely to feel devastated by setbacks or overly elated by praise, creating a stable foundation for consistent effort. Each deliberate action strengthens skill, confidence, and resilience. Momentum grows as effort continues steadily, independent of fluctuating results or external recognition. Smart students develop balance, focus, and sustained drive by valuing growth over judgment.

Separating self-worth from performance also improves focus on the learning process itself. Students who tie identity to outcomes may rush tasks or avoid deep engagement to ensure favorable results. This behavior limits understanding and skill development. By valuing effort and learning rather than external success, students can focus fully on the process. Each step taken with attention and intentionality builds competence and confidence. Momentum grows as mastery emerges from deliberate practice rather than the pursuit of validation.

Another benefit is reduced fear of feedback. When self-esteem is linked to results, criticism can feel threatening and discourage further effort. Students may ignore constructive feedback or avoid opportunities for growth. By detaching self-worth from outcomes, students can receive critique objectively and use it to improve. Each adjustment informed by feedback strengthens skill and reinforces autonomy. Momentum is reinforced as students act on insights without being hindered by emotional reactions. Learning becomes iterative and purposeful.

Detaching identity from achievement also enhances resilience in the face of setbacks. Students are less likely to interpret mistakes as personal failure and more likely to see them as opportunities for growth. This mindset reduces discouragement and fosters perseverance. Each challenge overcome without tying worth to success builds confidence and capability. Momentum grows as students continue taking action despite obstacles or imperfect outcomes. Growth becomes sustainable because effort is not penalized by temporary setbacks.

This approach also encourages authentic goal-setting. Students can pursue objectives aligned with personal values rather than those dictated by external expectations. This ensures that energy is invested in meaningful challenges that foster genuine development. Each step toward self-defined goals builds skill, competence, and confidence. Momentum is maintained as action is guided by purpose and intrinsic motivation rather than external judgment. Smart students achieve more when they act from alignment with personal priorities.

Separating self-worth from performance also strengthens long-term motivation. When students no longer rely on external outcomes to feel valuable, they can sustain effort even during periods of slow progress. This mindset encourages consistency, curiosity, and a love of learning for its own sake. Each action taken for growth, rather than validation, reinforces competence and self-trust. Momentum compounds as effort becomes habitual, guided by intrinsic motivation rather than fluctuating results. Smart students develop endurance and persistence, achieving meaningful growth over time.

Another benefit is the freedom to explore and innovate. Students who are not bound to performance-based self-esteem can take creative risks and try unconventional approaches. This experimentation fosters skill development, problem-solving, and adaptability. Each attempt, whether successful or not, provides valuable insight and builds confidence. Momentum grows as students act boldly without fear of judgment. By focusing on exploration over external approval, smart students cultivate both competence and resilience.

This mindset also improves emotional well-being. Students experience fewer extremes of elation or despair based on outcomes, reducing stress and enhancing focus. They are better able to manage setbacks and approach challenges with calm determination. Each step taken with balanced perspective strengthens confidence and self-trust. Momentum is maintained as effort continues steadily, independent of external rewards or recognition. Emotional stability allows smart students to sustain progress over the long term.

Separating self-worth from performance fosters a growth-oriented mindset. Students focus on learning, improvement, and skill acquisition rather than comparing themselves to others or achieving perfection. Each incremental gain reinforces competence and the ability to take further action. Momentum is created when consistent effort and reflection replace anxiety over results. Smart students thrive when growth, curiosity, and effort define their identity rather than outcomes.

Finally, cultivating self-worth independent of performance requires deliberate awareness and practice. Students must consciously reinforce their value based on effort, learning, and intrinsic goals. Each intentional action strengthens resilience, confidence, and momentum. Momentum is sustained when effort is guided by personal growth rather than external judgment. By separating self-worth from performance, smart students unlock authentic potential, take purposeful risks, and achieve meaningful, lasting results.

QUOTES:

"Your value doesn't decrease based on someone's inability to see your worth." — Unknown

Chapter .19

Learning to Think in Experiments Instead of Outcomes

Many smart students get stuck because they focus too heavily on achieving perfect results. This outcome-oriented thinking can create fear, hesitation, and overanalysis. By reframing tasks as experiments, students shift the focus from immediate success to learning and discovery. Experiments encourage iteration, feedback, and adaptation, allowing students to act without fear of failure. Each attempt provides valuable information, regardless of the result. Momentum builds as learning becomes the primary goal rather than flawless performance.

One reason students fixate on outcomes is past reinforcement from school and society. Praise and rewards often emphasize correct answers, high grades, or visible success. Students internalize the belief that their value depends on achieving perfect results. This mindset discourages risk-taking, creativity, and exploration. Thinking in experiments allows students to embrace uncertainty and see mistakes as opportunities. Each small test or trial contributes to skill development and confidence.

Another consequence of outcome-focused thinking is procrastination. Students may delay action, fearing that results won't meet expectations. This inaction reduces learning and prevents progress. By approaching tasks as experiments, students are free to start, test, and iterate without pressure. Each trial provides insight, reinforces learning, and strengthens problem-solving ability. Momentum grows as movement replaces hesitation and perfectionism.

Thinking in experiments also fosters adaptability. When students focus on learning rather than fixed outcomes, they can adjust strategies, test assumptions, and pivot based on results. This flexibility strengthens resilience, creativity, and skill acquisition. Each experiment, regardless of success, provides feedback that informs future decisions. Momentum is created as students continuously refine their approach through action and observation. Experimentation becomes a tool for consistent progress.

Thinking in experiments also reduces the fear of failure that often paralyzes smart students. When outcomes are treated as tests rather than judgments, mistakes become opportunities for learning. Students can analyze what worked, what didn't, and adjust their approach without self-criticism. Each experiment builds understanding and reinforces competence. Momentum grows as repeated testing leads to incremental improvements and confidence in one's abilities. This approach transforms fear into curiosity and action.

Another benefit is improved decision-making. Outcome-oriented students may overanalyze or delay choices, fearing negative results. Treating decisions as experiments allows students to take action while gathering feedback for refinement. Each small test informs future strategies, reducing uncertainty and enhancing judgment. Momentum builds as students act deliberately, learn from results, and adjust course efficiently. Thinking experimentally encourages practical learning over endless deliberation.

Approaching challenges as experiments also fosters innovation. Students feel freer to explore unconventional solutions without the pressure of immediate success. Creative ideas can be tested, refined, or discarded based on evidence rather than fear of judgment. Each trial enhances problem-solving skills and broadens perspectives. Momentum grows as students iterate quickly, learning from every attempt. An experimental mindset unlocks both creativity and practical competence.

Experiments also reinforce resilience and persistence. When outcomes are not tied to personal worth, setbacks are less discouraging. Students can continue testing, learning, and adjusting without fear of failure. Each iteration strengthens understanding, skill, and confidence. Momentum is maintained as effort becomes continuous, informed by observation rather than inhibited by perfectionism. Smart students learn that progress comes from repeated action, not from flawless results.

Experiments also improve efficiency in learning and problem-solving. Outcome-focused students often spend excessive time trying to guarantee success before taking action, which slows progress. By testing ideas in small, manageable ways, students can gather feedback quickly and adjust their approach. Each experiment provides valuable data that informs future decisions and prevents wasted effort. Momentum grows as repeated testing accelerates understanding and skill acquisition. Thinking experimentally turns trial and error into a structured, productive process rather than a source of frustration.

Another benefit is reduced anxiety and mental pressure. When students fixate on achieving perfect results, stress and self-doubt can accumulate, undermining performance. Treating tasks as experiments shifts focus from judgment to learning, easing emotional burden. Each trial becomes an opportunity to learn rather than a measure of worth. Momentum builds as students act with curiosity instead of fear, allowing consistent progress. This mindset creates a safer, more productive learning environment for growth and exploration.

An experimental approach also fosters better reflection. Students can objectively analyze outcomes, identify patterns, and adjust strategies accordingly. Instead of reacting emotionally to setbacks, they gather information and apply it to the next iteration. Each experiment strengthens problem-solving skills and deepens understanding. Momentum grows as insights from each trial inform subsequent actions, creating a feedback loop of continuous improvement. Smart students gain both knowledge and confidence by systematically learning from experience.

This mindset also encourages strategic risk-taking. Students become more willing to attempt novel approaches because failure is reframed as a source of information rather than a personal deficit. Each calculated risk provides feedback, expands capability, and strengthens adaptability. Momentum is maintained as action continues despite uncertainty or imperfect results. Thinking in experiments allows smart students to explore new possibilities while steadily building competence. This approach converts curiosity and creativity into tangible progress.

Thinking in experiments also strengthens problem-solving skills over time. When students focus on learning rather than immediate success, they develop the ability to test multiple approaches and find what works best. Each experiment, regardless of outcome, provides insight into strategies, methods, and personal strengths. Momentum builds as repeated trials lead to refinement and increasingly effective solutions. This approach trains smart students to approach challenges analytically and creatively. Learning becomes an active, iterative process rather than a passive pursuit of correct answers.

Another benefit is the cultivation of patience and persistence. Outcome-driven students may become frustrated when results are not immediate, which can reduce effort or motivation. By viewing tasks as experiments, they understand that learning takes time and incremental improvement is valuable. Each small test reinforces progress, even if it does not produce perfect results. Momentum is sustained as students focus on consistent action rather than instant achievement. Patience becomes a skill, supporting long-term growth and resilience.

Adopting an experimental mindset also reduces fear of uncertainty. Many smart students hesitate to act when the outcome is unknown, fearing mistakes or judgment. Thinking in experiments reframes uncertainty as a necessary and valuable component of learning. Each trial, whether successful or not, adds information and builds confidence. Momentum grows as students embrace ambiguity as a tool for growth rather than a barrier. This perspective allows smart students to act boldly, explore creatively, and continuously improve.

Experiments encourage adaptability and continuous improvement. Students can adjust methods based on observations and results, creating a cycle of learning and refinement. This process ensures that effort is always moving toward growth, not just busy work. Each iteration reinforces skill, competence, and understanding. Momentum is maintained as students respond dynamically to feedback rather than being hindered by perfectionism or fear. The experimental mindset creates a resilient and resourceful approach to learning and problem-solving.

Thinking in experiments also fosters self-confidence in the learning process. Students begin to trust that taking action, observing results, and adjusting accordingly will lead to improvement over time. This trust reduces hesitation and encourages initiative, even when outcomes are uncertain. Each trial reinforces competence and the ability to navigate challenges effectively. Momentum grows as students act consistently, knowing that progress comes from iterative learning rather than immediate perfection. Over time, this mindset strengthens both skill and self-assurance.

Another advantage is enhanced creativity and innovation. Students who see tasks as experiments feel freer to explore unconventional approaches without fear of failure. Each test, whether successful or not, provides new insights and inspires alternative strategies. Momentum builds as experimentation encourages continual exploration and idea generation. This iterative process allows smart students to develop novel solutions and creative thinking skills. Thinking experimentally transforms curiosity into actionable progress.

This mindset also improves resilience and stress management. Students are less likely to internalize mistakes as personal failure, reducing anxiety and self-doubt. Instead, setbacks become valuable feedback for refining strategies. Each experiment, even if it does not yield perfect results, strengthens adaptability and emotional endurance. Momentum grows as students learn to persist, adjust, and continue acting despite challenges. Resilience becomes a habit, supporting long-term growth and consistent performance.

Adopting an experimental mindset encourages a practical approach to learning. Students learn to break complex problems into smaller, testable components. Each small trial provides insight and builds understanding, making larger challenges more manageable. Momentum is reinforced as consistent action produces cumulative results over time. Smart students develop a systematic approach to learning, where curiosity, reflection, and iteration create steady progress. This transforms intelligence into practical, achievable outcomes.

Finally, cultivating an experimental mindset requires conscious effort and reflection. Students must intentionally reframe tasks as opportunities to test, learn, and iterate. Each small experiment builds knowledge, competence, and confidence. Momentum grows as repeated action produces measurable improvement and insight. By thinking in experiments instead of outcomes, smart students unlock creativity, resilience, and continuous growth while transforming intelligence into tangible results.

QUOTES:

"Fail often so you can succeed sooner."

— Tom Kelley

Chapter .20

Why Boring Consistency Beats Brilliant Bursts

Many smart students fall into the trap of relying on occasional bursts of intense effort rather than steady, consistent work. They may believe that moments of brilliance or last-minute inspiration are enough to achieve results. While occasional bursts can produce short-term gains, they rarely create lasting skill, understanding, or momentum. Consistency, even when boring or unremarkable, compounds over time and builds sustainable growth. Small, repeated actions strengthen habits, confidence, and competence. Momentum grows as consistent effort gradually accumulates into meaningful progress.

One reason consistency is so powerful is that it allows for repeated practice and reinforcement. Skills and knowledge improve incrementally through frequent engagement rather than sporadic effort. Students who act consistently are more likely to internalize concepts, master techniques, and develop durable habits. Each step, even if small, reinforces learning and builds confidence. Momentum emerges as steady effort transforms into expertise and reliable performance. Over time, consistent action yields results far beyond what occasional bursts can achieve.

Another benefit of boring consistency is reduced stress and burnout. Students who rely on intense, irregular effort often experience exhaustion, pressure, and anxiety. In contrast, steady, manageable effort balances workload and energy levels. Each day of deliberate, sustainable action contributes to progress without overwhelming mental or physical resources. Momentum is maintained because effort is continuous rather than erratic. Smart students who embrace consistency develop resilience and stamina for long-term achievement.

Consistency also fosters accountability and habit formation. Small, repeated actions become automatic over time, reducing the need for motivation or willpower. Students no longer rely on occasional inspiration or intense effort to make progress. Each day of consistent action reinforces routines and builds skill incrementally. Momentum grows as habits replace reliance on motivation, ensuring steady advancement toward goals. Boring consistency creates a reliable framework for success that bursts of brilliance cannot sustain.

Consistency also allows for better tracking and measurement of progress. When students act regularly, they can observe patterns, identify areas for improvement, and make data-driven adjustments. Sporadic bursts, no matter how intense, provide limited feedback and make it difficult to evaluate effectiveness. Each day of steady effort creates a record of growth and informs future action. Momentum builds as consistent practice generates clear evidence of improvement. Over time, students can refine strategies and optimize learning based on accumulated experience.

Another advantage is that consistency reduces the risk of procrastination. Students who wait for inspiration or the perfect moment to act often delay meaningful work. Regular, small actions create a routine that bypasses hesitation and builds discipline. Each repeated effort reinforces the habit of acting promptly, reducing the mental friction that leads to procrastination. Momentum grows as action becomes automatic rather than sporadic. Smart students learn that consistent steps, however small, produce more results than waiting for sporadic motivation.

Consistency also strengthens resilience and emotional endurance. Students who rely on occasional intense effort may face discouragement when results are delayed or insufficient. Steady effort builds confidence, self-trust, and patience. Each day of regular work reinforces the understanding that progress comes from accumulation, not sudden bursts. Momentum is maintained as students continue acting despite setbacks or slow initial results. Consistent practice creates a stable foundation for both achievement and personal growth.

This approach also enhances focus and reduces wasted energy. Sporadic bursts can be mentally draining, requiring intense concentration for short periods followed by fatigue and recovery. Boring consistency distributes effort evenly, allowing sustained focus and higher overall productivity. Each day of manageable action strengthens skills and understanding while conserving energy. Momentum grows as students act steadily, avoiding burnout and maintaining long-term engagement. Smart students achieve more by pacing themselves consistently rather than relying on occasional peaks.

Consistency also builds mastery through repetition. Skills, knowledge, and habits are strengthened when practiced regularly, even if the practice feels mundane or routine. Sporadic bursts may create temporary improvement but rarely lead to deep, long-lasting competence. Each day of consistent practice reinforces neural pathways, making actions more efficient and intuitive over time. Momentum grows as repeated effort compounds into expertise and reliability. Smart students learn that mastery is built quietly, step by step, rather than through occasional displays of brilliance.

Another benefit is the reduction of performance pressure. When students rely on intense, sporadic effort, each attempt feels high-stakes and anxiety-inducing. Consistent, steady effort spreads the pressure over time and makes progress feel achievable. Each small, deliberate action builds skill and confidence without triggering overwhelming stress. Momentum is sustained because effort becomes routine and manageable, not an all-or-nothing endeavor. This approach fosters a healthier, more sustainable learning experience.

Consistency also encourages incremental improvement and refinement. Students can observe gradual changes, identify patterns, and adjust techniques or strategies as needed. Sporadic bursts provide little feedback and can create a false sense of progress. Each day of steady practice allows for evaluation, adaptation, and improvement. Momentum grows as students integrate lessons from daily effort into better performance. Over time, these small improvements accumulate into significant growth.

This approach also strengthens self-discipline and accountability. By committing to consistent action, students cultivate habits that reduce reliance on motivation or external pressure. Each day of deliberate work reinforces personal responsibility and follow-through. Momentum builds as habits replace the need for constant decision-making or bursts of effort. Smart students discover that reliability and discipline consistently produce better results than sporadic brilliance.

Consistency also enhances long-term motivation and persistence. When progress is built through steady action, students see the cumulative effect of repeated effort over time. This reinforces confidence and encourages continued work, even when immediate results are modest. Sporadic bursts, in contrast, often produce uneven outcomes and can lead to frustration or burnout. Each day of consistent effort strengthens the habit of persistence and reduces reliance on fleeting motivation. Momentum grows as steady action compounds into measurable, lasting achievement.

Another benefit is the development of focus and efficiency. Regular, predictable effort allows students to plan their time and energy strategically. They can concentrate on high-priority tasks and steadily advance toward goals without the distraction of irregular, high-intensity work. Each consistent action builds skill and understanding incrementally. Momentum is reinforced as small, deliberate steps accumulate into significant accomplishments. Smart students achieve more when effort is reliable rather than sporadic.

Consistency also supports emotional stability. Students who rely on bursts of brilliance often experience emotional highs and lows tied to results. Steady, repeated effort reduces stress, anxiety, and self-criticism by creating predictable progress. Each day of manageable action reinforces confidence and diminishes the fear of failure. Momentum is sustained as effort becomes habitual and emotionally sustainable. A stable, consistent approach allows smart students to maintain energy and focus over the long term.

This approach also strengthens long-term learning and retention. Skills and knowledge are more deeply embedded when practiced regularly, rather than in intense, isolated sessions. Each day of consistent review and application reinforces understanding and improves memory. Momentum grows as repeated practice transforms effort into lasting competence. Smart students find that boring consistency produces results that brilliance alone cannot achieve. Small, steady steps create durable, meaningful progress over time.

Consistency also builds cumulative advantage over time. Each small, repeated action compounds, creating skills, knowledge, and habits that multiply in value. Sporadic bursts may produce brief results, but they rarely generate long-term momentum or expertise. By acting steadily, students create a foundation that supports larger achievements and greater opportunities. Each day of consistent effort reinforces competence, confidence, and resilience. Momentum grows as small, daily actions accumulate into substantial, lasting results that would be impossible through occasional brilliance alone.

Another benefit is the reinforcement of self-trust. Students who consistently follow through on small commitments learn to rely on themselves. They see evidence that steady action produces results, which strengthens confidence in their own abilities. Each completed task, no matter how small, builds credibility and trust in one's capacity to deliver. Momentum is sustained as students internalize the reliability of their own effort. Smart students gain autonomy and assurance through consistent, disciplined action.

Consistency also improves adaptability and problem-solving. Regular practice exposes students to repeated challenges, requiring them to adjust strategies and refine techniques over time. Sporadic bursts provide little opportunity to learn from patterns or iterative experience. Each consistent action creates a feedback loop that informs future decisions and enhances skill. Momentum grows as students act, evaluate, and adapt continuously, turning routine effort into progressive mastery. This iterative approach ensures learning is both deep and flexible.

This mindset also fosters patience and long-term perspective. Students who rely on occasional bursts may become frustrated by slow progress or delayed results. Steady, incremental effort demonstrates that growth is a process, not an instant outcome. Each day of consistent action reinforces the understanding that mastery and achievement are built over time. Momentum is maintained as students develop endurance, patience, and a long-range view of their learning journey. Smart students learn to value the process as much as the results.

Finally, embracing boring consistency over brilliant bursts requires intentionality, discipline, and mindset shifts. Students must prioritize daily effort, track progress, and resist the temptation to rely on sporadic inspiration. Each small, repeated action strengthens competence, confidence, and momentum. Momentum is sustained as effort becomes habitual, predictable, and productive. By valuing boring consistency, smart students unlock sustainable growth, mastery, and meaningful results that compound over the long term.

QUOTES:

"Small disciplines repeated with consistency every day lead to great achievements gained slowly over time."

— John C. Maxwell

Chapter .21

Training Your Brain to Tolerate Discomfort

Many smart students avoid discomfort because it feels unpleasant and threatens their sense of competence. Challenges, mistakes, and uncertainty often trigger anxiety, leading to avoidance or procrastination. However, growth requires stepping into these uncomfortable situations consistently. By training the brain to tolerate discomfort, students can approach challenges with resilience, curiosity, and determination. Each experience of pushing through discomfort strengthens mental endurance and confidence. Momentum builds as repeated exposure to difficult tasks reduces fear and resistance over time.

One reason discomfort feels so threatening is that students often tie it to failure or inadequacy. Struggling with a task may feel like proof of incompetence, reinforcing self-doubt. This mindset discourages risk-taking, experimentation, and learning from mistakes. By separating discomfort from personal worth, students can engage fully in challenging tasks without fear. Each instance of enduring difficulty teaches that effort, not ease, produces growth. Momentum is created as discomfort becomes a signal for opportunity rather than a reason to retreat.

Another consequence of avoiding discomfort is the development of stagnation and complacency. When students consistently choose comfort over challenge, skill development slows, and potential remains untapped. Facing uncomfortable tasks regularly expands capacity, knowledge, and resilience. Each time a student tolerates discomfort, they expand their threshold for challenge. Momentum grows as students gradually take on increasingly complex and demanding tasks. Growth accelerates when discomfort is reframed as a necessary step toward mastery.

Training the brain to tolerate discomfort also improves emotional regulation. Students learn to stay focused, composed, and reflective even when tasks are difficult or outcomes uncertain. This reduces stress and anxiety while increasing perseverance. Each encounter with discomfort strengthens coping strategies and reinforces confidence in one's ability to handle challenges. Momentum is maintained as students continue acting effectively under pressure. Smart students develop a mindset where difficulty signals progress, not failure.

Repeated exposure to discomfort also builds confidence in one's abilities. When students face difficult tasks regularly, they learn that they can survive and even thrive despite uncertainty or challenge. Each experience reinforces the understanding that struggle is a natural part of growth. Momentum grows as students internalize their capacity to handle adversity. Over time, tasks that once felt intimidating become manageable. Smart students develop courage and self-trust through consistent practice in tolerating discomfort.

Another benefit is improved problem-solving and creativity. Discomfort often signals unfamiliar situations or challenges that require new approaches. When students engage rather than avoid, they are forced to think critically and adapt. Each encounter with difficulty provides opportunities to experiment, adjust strategies, and discover innovative solutions. Momentum builds as students become skilled at navigating complexity and uncertainty. The brain learns to associate challenge with opportunity rather than threat.

Training the brain to tolerate discomfort also strengthens perseverance. Students who avoid difficulty may give up when tasks feel challenging or results are slow. Embracing uncomfortable situations cultivates patience, endurance, and a long-term perspective. Each time students push through resistance, they reinforce habits of persistence. Momentum is maintained as effort continues even when progress is not immediate. This sustained action builds both competence and resilience.

This mindset also reduces the emotional impact of setbacks. Students who habitually face discomfort experience fewer negative emotional swings when results are imperfect. They learn to respond with reflection, adjustment, and persistence rather than frustration or self-criticism. Each exposure strengthens emotional regulation and builds resilience. Momentum grows as students act confidently under pressure, even when outcomes are uncertain. Smart students learn to separate discomfort from personal failure.

Gradually increasing exposure to discomfort also builds mental toughness. Students can start with small challenges that push them slightly out of their comfort zone and progressively take on more difficult tasks. This incremental approach prevents overwhelm while steadily strengthening resilience. Each successful step reinforces the belief that discomfort is manageable and valuable for growth. Momentum grows as students accumulate experiences that expand their capacity to face challenges. Smart students learn that enduring small difficulties prepares them to handle larger, more complex obstacles.

Another benefit is the development of patience and delayed gratification. Facing discomfort often requires sustained effort over time, rather than expecting immediate results. Students learn that progress comes through persistence, reflection, and incremental improvement. Each encounter with difficulty reinforces the value of consistent effort and long-term thinking. Momentum builds as students develop the habit of working steadily through challenges, rather than seeking quick fixes. This mindset cultivates discipline and fortitude in pursuit of meaningful goals.

Training the brain to tolerate discomfort also improves adaptability. Students who regularly confront challenges learn to adjust strategies, respond to unexpected obstacles, and recover from setbacks. Each uncomfortable situation becomes an opportunity to practice flexibility and problem-solving. Momentum grows as students develop the ability to pivot, innovate, and act decisively under pressure. This adaptability ensures sustained progress even in complex, unpredictable circumstances. Smart students transform challenges into learning opportunities through deliberate action.

Consistently facing discomfort also strengthens focus and engagement. Students learn to concentrate on the task at hand rather than avoiding it due to fear or discomfort. Each moment spent persisting despite difficulty enhances skill development, understanding, and confidence. Momentum is reinforced as students complete tasks that once felt intimidating or overwhelming. Over time, tolerance for discomfort becomes a habit, making even complex challenges approachable. This approach builds both competence and the resilience to maintain progress over time.

Training the brain to tolerate discomfort also strengthens emotional resilience. Students learn to manage frustration, stress, and uncertainty without being derailed. Each encounter with discomfort teaches coping strategies, patience, and self-regulation. Momentum grows as students continue acting despite emotional challenges, building both competence and confidence. Over time, situations that once caused stress become manageable learning opportunities. Smart students develop the emotional endurance needed to persist through demanding tasks and complex challenges.

Another benefit is improved decision-making under pressure. When students face uncomfortable situations, they must think critically and act despite uncertainty. Avoiding discomfort often leads to indecision or procrastination. Each time students confront difficulty, they practice evaluating options, taking calculated risks, and learning from outcomes. Momentum builds as confidence in decision-making grows, reducing hesitation in future challenges. This enhances both personal and professional effectiveness over time.

Training the brain to tolerate discomfort also fosters a growth-oriented mindset. Students come to view challenges as opportunities rather than threats. Each experience of navigating difficulty reinforces the understanding that effort, experimentation, and persistence lead to improvement. Momentum grows as students embrace challenges proactively rather than retreating to comfort zones. Over time, discomfort becomes a signal for learning and growth rather than a source of fear. Smart students thrive by leaning into challenges that expand skill and capability.

Consistent exposure to discomfort also builds self-reliance and autonomy. Students learn that they can navigate challenges independently, even when situations are unfamiliar or difficult. Each successful navigation reinforces trust in their own abilities and judgment. Momentum is maintained as students act confidently without waiting for external guidance or reassurance. This self-trust empowers them to take initiative, tackle complex problems, and maintain consistent progress. By embracing discomfort, smart students become more capable, independent, and resilient.

Over time, training the brain to tolerate discomfort transforms the way students approach challenges. What once felt intimidating or overwhelming becomes manageable and even motivating. Each experience builds not only skill but also the confidence to confront increasingly complex tasks. Momentum grows as students learn that discomfort is temporary and progress is cumulative. Smart students begin to anticipate challenges as opportunities to learn rather than threats to their competence. This shift in perspective fundamentally changes their approach to learning and growth.

Another result is the reduction of avoidance behaviors. Students who previously procrastinated or sidestepped difficult tasks now act despite discomfort. Each time they choose engagement over avoidance, they reinforce the habit of persistence and action. Momentum builds as consistent exposure to difficulty strengthens resilience and self-discipline. Over time, avoidance becomes less of a default response, and action becomes the norm. Smart students develop the habit of confronting challenges proactively rather than reactively.

Training the brain to tolerate discomfort also fosters a mindset of continuous improvement. Students learn that every difficult experience carries lessons that can be applied to future challenges. Each trial, mistake, or failure becomes data for refining strategies and building competence. Momentum grows as students iterate, adjust, and act with increasing effectiveness. Over time, discomfort is reframed as a tool for growth rather than a source of stress. Smart students approach learning as a process of experimentation and development.

This practice also strengthens patience and long-term perspective. Students understand that mastery and meaningful results require sustained effort over time. Each moment of enduring difficulty reinforces the value of persistence and incremental progress. Momentum is maintained as students act steadily, even when outcomes are delayed or imperfect. Over time, the cumulative effect of consistently facing discomfort produces growth that bursts of effort cannot replicate. Smart students cultivate endurance, focus, and resilience that support lifelong learning.

Finally, cultivating tolerance for discomfort requires deliberate attention, repetition, and reflection. Students must consciously choose to engage with challenges, analyze what each experience teaches, and continue acting despite difficulty. Each step strengthens competence, confidence, and momentum. Momentum is sustained as repeated action transforms discomfort into opportunity for growth. By training the brain to tolerate discomfort, smart students unlock resilience, adaptability, and the sustained drive necessary for continuous achievement and meaningful, long-term success.

QUOTES:

"Greatness begins beyond your comfort zone." — Robin Sharma

CHAPTER .22

REPLACING "SHOULD" WITH "NEXT"

Many smart students feel trapped by an endless list of obligations and expectations. Words like "should" dominate their thinking, creating pressure, guilt, and indecision. "I should study more," "I should start that project," or "I should be further ahead" can paralyze action. Replacing "should" with "next" shifts the focus from judgment to movement. Instead of being weighed down by what they think they must do, students focus on the very next actionable step. Momentum grows as attention turns to concrete actions rather than abstract expectations.

One reason "should" is so paralyzing is that it implies a moral judgment. Students may feel they are failing themselves, others, or their potential if they don't act perfectly. This pressure creates anxiety and reduces clarity, making it hard to take any action. By reframing tasks as "next" steps, students remove judgment and focus on immediate action. Each small step becomes manageable, achievable, and free from emotional weight. Momentum is reinforced as consistent action replaces guilt-driven hesitation. Smart students learn that progress comes from movement, not from moralizing expectations.

Another consequence of "should" thinking is procrastination. Students may endlessly plan, revise, or wait for motivation, believing they must meet some ideal standard. This mindset delays results and prevents learning from experience. Thinking in terms of "next" steps encourages starting with the smallest possible action toward a goal. Each completed action, no matter how minor, produces insight, experience, and momentum. Momentum grows as action becomes habitual, replacing hesitation and overthinking. By focusing on the next step, smart students turn planning into execution.

Replacing "should" with "next" also strengthens clarity and prioritization. Students no longer feel overwhelmed by the infinite list of tasks they believe they should do. Instead, they identify the most important or immediate action and commit to it. Each next step moves them closer to meaningful progress while keeping attention manageable. Momentum builds as tasks are addressed sequentially, creating a sense of accomplishment and forward motion. Smart students achieve more by acting deliberately rather than being stalled by obligation or perfectionism.

Focusing on the "next" step also reduces overwhelm and decision fatigue. When students view tasks through the lens of "should," the sheer number of obligations can feel paralyzing. Thinking in terms of "next" breaks big projects into manageable actions. Each small step feels achievable and avoids mental overload. Momentum grows as students accomplish concrete tasks rather than being stuck in contemplation. Smart students learn that clarity and progress come from simplifying action, not expanding pressure.

Another benefit is increased adaptability. Life is unpredictable, and circumstances often make it impossible to complete every "should." Students who operate from obligation may feel frustrated or guilty when plans are disrupted. Thinking in terms of "next" allows them to adjust, reprioritize, and take the next logical step regardless of circumstances. Each action, even if imperfect, contributes to momentum and progress. Momentum is reinforced as students respond fluidly to change instead of being rigidly tied to expectations. Flexibility becomes a habit, supporting long-term growth and resilience.

Replacing "should" with "next" also fosters a sense of empowerment. Students gain control over what they can actually do in the moment rather than being dominated by external or internal pressures. Each next step taken reinforces agency, autonomy, and confidence. Momentum grows as students realize that even small actions create forward motion. Over time, this approach replaces procrastination and guilt with proactive movement and practical problem-solving. Smart students thrive when they focus on actionable choices rather than abstract obligations.

This mindset also enhances focus and attention. Instead of scattering energy across an endless list of "shoulds," students concentrate on the immediate next action. Each step completed clarifies the path forward and reinforces the habit of deliberate progress. Momentum builds as attention is focused on what matters most right now, creating a rhythm of action. Smart students achieve more consistently when they channel energy into sequential steps rather than juggling overwhelming expectations. Clarity and focus become natural outcomes of this approach.

Another advantage of thinking in terms of "next" is the reduction of perfectionism. Students who operate under "should" often feel that every step must be flawless, which delays action and increases stress. Viewing tasks as the next step emphasizes progress over perfection. Each small, imperfect action contributes to learning, skill-building, and forward momentum. Momentum grows as students take action without waiting for ideal conditions or flawless execution. Over time, the habit of focusing on next steps replaces fear of imperfection with productive engagement.

This mindset also encourages consistent action over sporadic bursts. Students who think in terms of "should" may procrastinate until motivation or inspiration strikes, often leading to rushed, inconsistent work. Focusing on the next step promotes steady, manageable progress every day. Each action, no matter how small, compounds into measurable results over time. Momentum builds as repeated, intentional steps accumulate into larger accomplishments. Smart students achieve more through consistent movement rather than relying on occasional moments of intensity.

Replacing "should" with "next" also strengthens decision-making and prioritization skills. When overwhelmed by obligations, students often struggle to determine what to do first. By identifying the immediate next step, they simplify choices and reduce mental friction. Each completed step clarifies priorities for subsequent actions. Momentum grows as students move sequentially through tasks, maintaining forward progress without getting stuck in indecision. Over time, this approach trains the brain to focus on what is actionable and relevant rather than theoretical or abstract.

This mindset also fosters resilience in the face of setbacks. Students who operate under "should" often give up when plans do not go perfectly or when obstacles arise. Thinking in terms of "next" encourages adaptation and forward movement despite challenges. Each step taken, even under imperfect conditions, reinforces confidence and competence. Momentum is maintained as students continue progressing, learning from setbacks rather than being paralyzed by them. Smart students develop the habit of acting with flexibility, persistence, and purpose.

Focusing on "next" also enhances motivation by making progress tangible. When students are trapped in "should" thinking, goals can feel distant and overwhelming, reducing the drive to act. By concentrating on the immediate next step, progress becomes visible and achievable. Each small action produces a sense of accomplishment and reinforces momentum. Momentum grows as repeated next steps accumulate into measurable advancement. Smart students maintain motivation by celebrating movement rather than fixating on abstract obligations.

Another benefit is the reduction of mental clutter and stress. "Should" thinking often creates a constant mental list of tasks, obligations, and expectations, which can feel chaotic and paralyzing. Thinking in terms of next steps simplifies focus to one actionable item at a time. Each completed step clears mental space and creates clarity for the following action. Momentum is maintained as energy is directed toward tangible progress rather than scattered across overwhelming demands. Over time, the brain learns to prioritize movement over worry.

This mindset also strengthens follow-through and accountability. Students who focus on next steps develop a habit of taking immediate, intentional action rather than delaying due to indecision or fear. Each action reinforces reliability and personal responsibility. Momentum grows as the habit of executing small steps daily compounds into larger achievements. Smart students build self-trust by consistently doing what is within their control instead of being paralyzed by obligations. Action becomes the engine of progress.

Replacing "should" with "next" also improves clarity in goal-setting. Students learn to break larger objectives into sequential, actionable steps. Each next step provides direction, making even complex goals more approachable. Momentum builds as students steadily move toward their aims, one deliberate action at a time. Over time, this approach transforms overwhelming ambitions into practical, achievable outcomes. Smart students discover that clarity and progress come from sequential action, not from guilt-driven pressure.

Over time, consistently focusing on the "next" step rewires the brain for action. Students begin to naturally prioritize immediate, meaningful tasks rather than ruminating over what they "should" do. Each completed next step reinforces a sense of agency and forward momentum. Momentum grows as taking action becomes the default response instead of hesitation or guilt. This habit transforms the learning process into a series of manageable, productive steps rather than an overwhelming list of obligations. Smart students learn that progress is created through movement, not judgment.

Another result is the strengthening of resilience and adaptability. When unexpected challenges or changes occur, students can identify the next logical action and proceed without being paralyzed by disappointment or frustration. Each small step in the face of difficulty reinforces confidence and problem-solving ability. Momentum is maintained as students continue moving forward, even when conditions are imperfect or uncertain. Over time, flexibility and persistence become automatic, allowing sustained growth. Smart students thrive by acting consistently and adapting fluidly to challenges.

Replacing "should" with "next" also encourages a mindset of experimentation. Instead of waiting for perfect conditions or fearing failure, students take actionable steps, learn from outcomes, and adjust accordingly. Each step provides feedback that informs subsequent action. Momentum grows as experience accumulates, and students refine their strategies through real-world learning. This approach transforms hesitation into productive movement and strengthens decision-making over time. Smart students understand that progress comes from doing, testing, and iterating, not from judgment or guilt.

This mindset also promotes self-compassion and reduces perfectionism. Students no longer equate missed "shoulds" with personal failure. Each next step is an opportunity to move forward, regardless of past mistakes or imperfections. Momentum builds as students focus on action rather than self-criticism. Over time, they develop a healthy relationship with their goals, seeing effort and progress as valuable in themselves. Smart students cultivate resilience and sustained motivation by prioritizing next steps over judgment.

Finally, adopting the "next" mindset requires ongoing intention and practice. Students must continually notice "should" thinking, convert obligations into actionable next steps, and follow through consistently. Each completed next step reinforces confidence, competence, and momentum. Momentum is maintained as action becomes habitual, manageable, and purposeful. By replacing "should" with "next," smart students transform stress, indecision, and perfectionism into clarity, forward movement, and continuous progress toward meaningful goals.

QUOTES:

"Start where you are. Use what you have. Do what you can." — Arthur Ashe

Chapter .23

How to Make Decisions Without Overthinking Them

Many smart students get stuck because they overanalyze every choice, fearing mistakes or imperfect outcomes. This overthinking creates paralysis, indecision, and wasted mental energy. Learning to make decisions without excessive rumination is a critical skill for maintaining momentum and achieving goals. By simplifying the decision process and focusing on actionable options, students can move forward with confidence. Each decision, even if imperfect, provides experience and insight for future choices. Momentum grows as repeated action replaces hesitation and mental clutter..

One reason overthinking is so common is that smart students often tie their identity to outcomes. They fear that making the wrong choice will reflect poorly on their intelligence, competence, or potential. This creates high-stakes pressure that slows action and undermines learning. By separating self-worth from the correctness of decisions, students can act more freely. Each decision becomes a step in experimentation rather than a judgment of personal value. Momentum is reinforced as action accumulates, even when outcomes are uncertain. Smart students learn that decisions are tools for learning, not measures of worth.

Another consequence of overthinking is procrastination. Students may delay action while considering every possible variable or outcome. This prevents progress and limits skill development. Learning to make decisions efficiently allows students to take immediate, purposeful action. Each choice, even a small one, produces information, feedback, and momentum. Momentum grows as students replace indecision with deliberate, timely movement. Smart students realize that progress comes from acting, not endlessly analyzing.

Simplifying the decision-making process also reduces mental fatigue. Constantly evaluating possibilities drains energy, reduces focus, and impairs judgment. By adopting practical strategies—such as limiting options, setting time limits, or prioritizing key criteria—students conserve cognitive resources. Each decision made efficiently frees energy for execution and learning. Momentum is maintained as mental clarity and action reinforce each other. Over time, smart students can make better choices faster while sustaining productivity.

Another benefit of making decisions without overthinking is increased confidence. Each time a student acts decisively, they reinforce the belief that they are capable of navigating uncertainty. Even imperfect choices provide valuable feedback that improves future decision-making. Momentum grows as repeated action builds both competence and trust in one's judgment. Over time, students learn that decisive movement often produces better results than endless deliberation. Smart students cultivate self-assurance by prioritizing action over hesitation.

This approach also reduces stress and anxiety. Overanalyzing decisions creates tension, worry, and a sense of being trapped. By simplifying choices and focusing on the next actionable step, students free themselves from unnecessary mental strain. Each decision taken produces clarity, progress, and reduced emotional burden. Momentum is maintained as students act with purpose rather than becoming stuck in cycles of doubt. Smart students develop calm, focused decision-making skills that support sustained growth.

Making decisions without overthinking also encourages experimentation and learning. Students learn to treat each choice as a hypothesis to test rather than a final judgment. Each decision, regardless of outcome, provides insights that inform future actions. Momentum builds as students refine strategies and develop practical problem-solving skills. Over time, this mindset transforms decision-making into a tool for growth rather than a source of fear. Smart students learn from experience rather than being paralyzed by the unknown.

This mindset strengthens efficiency and productivity. Students who overanalyze often spend excessive time weighing options, delaying action, and missing opportunities. By streamlining decision-making, they act faster and accomplish more with less mental energy. Each completed decision produces results and informs the next step. Momentum grows as consistent, timely choices accumulate into meaningful progress. Smart students achieve more by acting decisively rather than endlessly deliberating.

Another advantage of making decisions without overthinking is that it cultivates adaptability. Students who act decisively learn to adjust quickly when outcomes differ from expectations. Each choice, even if imperfect, provides information that informs the next step. Momentum grows as students respond fluidly to challenges rather than being stuck in indecision. Over time, this approach builds resilience and flexibility, allowing smart students to navigate complex or changing situations with confidence. The ability to pivot becomes a core strength that supports continuous learning and progress.

This mindset also fosters a sense of empowerment. Students who overthink often feel controlled by circumstances, fear, or expectations. By making deliberate decisions, they reclaim agency and control over their actions. Each choice reinforces autonomy and strengthens self-trust. Momentum is maintained as students see that they can act effectively despite uncertainty. Smart students gain the confidence to take initiative and shape their own learning and growth trajectories.

Making decisions efficiently also promotes clarity of focus. Overthinking can scatter attention across too many options, creating mental clutter. Focusing on actionable choices helps students concentrate on what matters most in the present moment. Each decision narrows focus, guiding energy toward meaningful action. Momentum grows as repeated focused choices accumulate into measurable progress. Smart students discover that clarity and action are more powerful than endless contemplation.

This approach encourages iterative learning. Students who act quickly can evaluate outcomes, adjust strategies, and make improved decisions in the future. Each decision serves as a learning opportunity that refines judgment and skill. Momentum builds as students integrate lessons from each action into subsequent choices. Over time, decision-making becomes faster, more confident, and increasingly effective. Smart students transform intelligence into practical, continuous improvement through action.

Another benefit of making decisions without overthinking is the reduction of opportunity cost. Students who delay action while analyzing endlessly often miss chances to learn, grow, or achieve. Each decisive step, even if imperfect, opens doors to new experiences and insights. Momentum grows as students act on opportunities rather than letting them slip by. Over time, the accumulation of timely decisions produces far more progress than waiting for the "perfect" choice. Smart students maximize learning and results by prioritizing action over indecision.

This mindset also strengthens problem-solving under pressure. Real-world challenges rarely wait for perfect clarity or full information. Students who practice making decisions efficiently learn to act effectively even when conditions are uncertain. Each choice, followed by reflection and adjustment, enhances adaptability and strategic thinking. Momentum builds as repeated action under pressure improves judgment and confidence. Over time, students develop the skill of navigating complex situations quickly and effectively. Smart students turn constraints and uncertainty into catalysts for growth.

Making decisions without overthinking also encourages experimentation and creativity. Students are more willing to try new approaches when they are not frozen by fear of making a wrong choice. Each decision becomes a test, providing insight into what works and what doesn't. Momentum grows as students iterate, learn from results, and refine strategies. Over time, experimentation becomes a natural part of learning and problem-solving. Smart students expand their capabilities by acting decisively and reflecting continuously.

This approach also fosters emotional resilience. Students who overanalyze often experience frustration, self-doubt, and anxiety when outcomes are not perfect. By taking action without overthinking, they learn to cope with imperfection and uncertainty. Each decision taken, regardless of result, strengthens confidence and emotional endurance. Momentum is maintained as students continue moving forward despite challenges or unexpected outcomes. Over time, they develop a mindset where action, learning, and adaptation are more important than worry or fear.

Finally, mastering the skill of making decisions without overthinking reinforces long-term growth and achievement. Students learn that progress is the result of repeated, deliberate action rather than endless deliberation. Each decision, regardless of perfection, contributes to experience, knowledge, and momentum. Momentum grows as the habit of decisive action compounds into meaningful accomplishments over time. Over time, smart students replace hesitation and self-doubt with confidence and forward movement. By acting decisively, they transform intelligence into sustained, practical results that continuously advance their goals.

Another important outcome is the development of trust in one's judgment. Students who consistently make decisions learn to rely on their own reasoning, intuition, and experience. Each successful step reinforces self-trust, while each misstep provides lessons without judgment. Momentum builds as students grow confident in their ability to handle uncertainty and adapt when outcomes are imperfect. Over time, trust in personal judgment strengthens decision-making speed, effectiveness, and independence. Smart students cultivate a balance of confidence, reflection, and action that empowers continuous progress.

This mindset also reduces mental fatigue and stress. Overanalyzing every choice consumes mental energy and drains motivation. By focusing on the next actionable step, students conserve cognitive resources while maintaining forward motion. Each decisive action alleviates mental clutter and reinforces productivity. Momentum is sustained as students replace hesitation with habit-driven movement. Over time, this creates a more efficient and sustainable approach to learning and growth.

Making decisions without overthinking also fosters creativity and adaptability. Students learn to act in uncertain situations, experiment, and adjust strategies as needed. Each choice provides feedback that informs future decisions and encourages innovative problem-solving. Momentum grows as students iterate quickly, respond to challenges, and leverage lessons learned. Over time, smart students develop the ability to create solutions under pressure rather than being paralyzed by analysis. This transforms intelligence into flexible, actionable capability.

Finally, cultivating this skill requires intentional practice and mindfulness. Students must notice when overthinking dominates, simplify options, and commit to immediate action. Each decision reinforces competence, confidence, and momentum. Momentum is sustained as decisive action becomes habitual, replacing hesitation and indecision with forward movement. By learning to make decisions without overthinking, smart students convert mental paralysis into continuous progress, turning their intelligence into consistent, meaningful, and measurable results.

QUOTES:

"Take action. An inch of movement will bring you closer to your goals than a mile of intention."

— Steve Maraboli

Chapter .24

Building Confidence Through Evidence, Not Affirmations

Many smart students struggle with confidence because they rely on positive affirmations rather than concrete evidence of their abilities. Saying "I am capable" or "I can do this" may feel motivating temporarily, but without proof of competence, confidence remains fragile. True confidence comes from repeated action, measurable progress, and accumulated experience. By focusing on evidence—completed tasks, demonstrated skills, and tangible results—students build a lasting sense of self-assurance. Each accomplishment reinforces capability and reduces doubt. Momentum grows as confidence becomes rooted in reality rather than wishful thinking.

One reason affirmations alone are ineffective is that they bypass the need for action. Students may repeat positive statements while avoiding the challenges that develop real skill. This creates a disconnect between perception and reality, leaving confidence shallow and easily shaken. By engaging in meaningful work and tracking results, students create evidence that supports genuine self-belief. Each task completed, skill learned, or problem solved provides tangible proof of competence. Momentum is reinforced as repeated action validates ability and strengthens trust in oneself. Smart students discover that confidence grows from doing, not merely thinking.

Another consequence of relying on affirmations is overconfidence or misplaced expectations. Without evidence, students may believe they are ready for challenges they have not actually practiced. This can lead to mistakes, frustration, or discouragement. Building confidence through evidence ensures readiness and competence in real situations. Each successful demonstration of skill, no matter how small, confirms the ability to handle challenges effectively. Momentum grows as students accumulate proof of competence, gradually increasing both confidence and capability. Smart students learn that measured, evidence-based confidence supports sustainable progress.

Evidence-based confidence also reduces anxiety and fear of failure. Students who trust that their skills are proven through action are less likely to be paralyzed by self-doubt. Each step, success, or completed task reinforces the belief that challenges can be met. Momentum is maintained as students act with assurance rather than hesitation. Over time, repeated evidence builds resilience, making confidence stable even in unfamiliar or difficult situations. Smart students transform competence into confidence by letting proof, not empty words, guide their belief in themselves.

Tracking evidence also helps students identify areas for growth. Confidence built solely on affirmations can overlook gaps in skill or understanding, leaving students unprepared for real challenges. By reviewing actual accomplishments and outcomes, students can see both strengths and areas needing improvement. Each identified improvement becomes an actionable next step that reinforces learning and capability. Momentum grows as students address gaps, turning evidence into both confidence and skill development. Smart students develop a realistic, actionable sense of self-assurance that propels them forward.

Another benefit of evidence-based confidence is that it encourages consistent action. Students who rely on affirmations alone may wait to "feel confident" before acting, leading to procrastination. When confidence is grounded in proven competence, students are more willing to take on challenges immediately. Each completed task reinforces their ability to handle more complex work. Momentum is maintained as action reinforces evidence, which in turn strengthens confidence. Over time, this creates a virtuous cycle of competence, confidence, and continuous progress.

Confidence built on evidence also reduces susceptibility to external validation. Students who rely on praise or recognition from others often tie self-worth to approval. Evidence-based confidence is internally grounded, stemming from demonstrated ability rather than external judgment. Each task completed, problem solved, or skill mastered strengthens autonomy and self-trust. Momentum grows as students rely on their own proven capability rather than fluctuating external opinions. Smart students cultivate resilience and independence through confidence grounded in reality.

This approach also strengthens decision-making and risk-taking. Students who know they have demonstrated competence are more likely to take calculated risks and seize opportunities. Each successful action reinforces the belief that they can handle uncertainty effectively. Momentum builds as repeated experiences validate skill and judgment. Over time, confidence becomes a reliable foundation for growth, enabling students to tackle increasingly challenging situations. Smart students transform evidence into empowerment, creating a mindset that balances caution with courage.

Evidence-based confidence also fosters perseverance in the face of setbacks. Students who rely solely on affirmations may feel discouraged when challenges arise or results fall short. Confidence grounded in evidence reminds them of past successes and proven capabilities, providing motivation to persist. Each obstacle overcome reinforces the knowledge that effort and skill produce results. Momentum grows as students continue acting despite difficulties, drawing strength from real accomplishments. Smart students build resilience by letting tangible evidence guide their belief in what they can achieve.

Another advantage is improved focus and clarity. When students trust in their proven abilities, they can concentrate on the task at hand rather than doubting their competence. Each action is guided by the confidence of prior achievements rather than vague self-assurances. Momentum builds as students tackle challenges with clear direction and purposeful effort. Over time, evidence-based confidence creates a sense of certainty in ability, reducing hesitation and second-guessing. Smart students maximize productivity by letting demonstrated skill guide action rather than relying on wishful thinking.

This approach also promotes humility and realistic self-assessment. Confidence built on evidence encourages students to recognize both strengths and limitations. Each success is acknowledged, and each gap becomes an opportunity for growth. Momentum grows as students continuously refine skills while building confidence based on real achievement. Over time, this balance prevents overconfidence while fostering a mindset oriented toward improvement and learning. Smart students integrate competence, reflection, and self-awareness for sustainable growth.

Evidence-based confidence also strengthens collaboration and influence. Students who are grounded in proven ability can contribute meaningfully in teams and decision-making situations. Each demonstration of competence reinforces credibility and trust with peers, mentors, and collaborators. Momentum builds as evidence of capability supports both individual and collective progress. Over time, students learn to leverage their strengths confidently while remaining open to learning from others. Smart students transform their evidence-based competence into effective action and positive impact.

Confidence built on evidence also encourages calculated risk-taking. Students who trust their proven abilities are more willing to step into unfamiliar or challenging situations. Each successful action reinforces the belief that they can handle complexity and uncertainty. Momentum grows as students experiment, learn, and refine strategies based on real results. Over time, this approach strengthens resilience and problem-solving capacity. Smart students transform evidence into courage, enabling them to take action rather than remain paralyzed by doubt.

Another benefit is the development of intrinsic motivation. Confidence based on accomplishments drives students to continue learning and improving for its own sake rather than seeking external validation. Each completed task and demonstrated skill reinforces the value of effort and growth. Momentum builds as repeated success fuels a desire for further mastery and achievement. Over time, intrinsic motivation becomes self-sustaining, reducing reliance on praise, affirmations, or temporary boosts. Smart students cultivate a deep, self-reinforcing drive through evidence-based confidence.

Evidence-based confidence also enhances goal-setting and planning. Students who recognize their capabilities can set realistic, challenging, and achievable goals. Each goal is informed by prior successes, making it both motivating and attainable. Momentum grows as students act consistently toward these objectives, using evidence to guide progress. Over time, they develop a clear understanding of what they can achieve and how to structure steps to get there. Smart students turn proven competence into purposeful, goal-oriented action.

This approach also strengthens resilience to criticism and external judgment. Students grounded in evidence are less shaken by negative feedback because they can rely on proof of their competence. Each completed task and skill demonstrated provides a foundation to interpret criticism constructively rather than personally. Momentum is maintained as students continue learning, improving, and taking action despite setbacks or doubts from others. Over time, confidence becomes independent of external opinions and firmly rooted in demonstrated ability. Smart students navigate challenges with poise because their confidence is grounded in reality.

Finally, confidence built through evidence fosters long-term resilience and independence. Students no longer need constant reassurance or motivation from external sources, because they can rely on proven results to guide their actions. Each accomplishment, no matter how small, reinforces their capability to handle future challenges. Momentum grows as repeated successes compound, creating a steady, reliable foundation of self-assurance. Over time, students internalize the understanding that competence leads to confidence, and confidence fuels further competence. Smart students transform intelligence into action, growth, and sustainable achievement by letting evidence—not empty words—shape their self-belief.

Another key outcome is improved decision-making. When students trust their demonstrated abilities, they can approach choices with clarity and confidence. Each decision made with evidence-based confidence is more deliberate, informed, and effective. Momentum is maintained as repeated successes reinforce the ability to make sound judgments under pressure. Over time, this strengthens both competence and agency, allowing students to tackle increasingly complex challenges. Smart students leverage proven skill as a guide for intelligent, confident action.

Evidence-based confidence also enhances persistence and grit. Students who understand their abilities are more likely to persevere when tasks are difficult or outcomes are uncertain. Each challenge faced and navigated successfully adds to a growing foundation of proof and experience. Momentum builds as perseverance becomes habitual, supported by tangible evidence of capability. Over time, setbacks are reframed as opportunities to demonstrate skill and learn. Smart students cultivate resilience by letting real achievements, not self-talk alone, fuel their persistence.

This approach also fosters a realistic sense of self-awareness. Students grounded in evidence recognize both strengths and areas for growth, creating a balanced and actionable perspective. Each reflection on accomplishments informs future goals and strategies. Momentum grows as students continuously apply lessons from past achievements to new challenges. Over time, they develop a practical, adaptable approach to learning and performance. Smart students combine confidence, competence, and reflection to sustain meaningful progress and achievement.

QUOTES:

"Action is the foundational key to all success."
— Pablo Picasso

Part IV: Turning Intelligence Into Momentum

Chapter .25

Designing Systems That Carry You Forward

Many smart students rely too heavily on willpower and motivation to get things done, only to find themselves stuck when energy or focus wanes. Systems, not sheer effort, create consistent progress and reduce mental friction. By designing structures, routines, and habits that guide action, students can ensure forward movement even when motivation is low. Each system acts like a conveyor belt, carrying them toward goals with minimal reliance on willpower alone. Momentum grows as repeated actions are built into reliable, automatic frameworks. Smart students transform intelligence into consistent results by leveraging systems rather than temporary energy or inspiration.

One reason systems are so effective is that they reduce decision fatigue. Students often waste energy deciding what to do next, which slows progress and drains focus. A well-designed system provides clarity, direction, and priority, allowing attention to be applied where it matters most. Each step taken within a system reinforces habits, skill, and confidence. Momentum is maintained as students move efficiently through tasks without getting bogged down by constant choices. Over time, systems allow intelligence to be translated into consistent, high-quality action.

Another consequence of ignoring systems is reliance on motivation, which is unpredictable and fleeting. Students who wait for inspiration often experience bursts of productivity followed by long periods of stagnation. Systems create a dependable structure that produces progress regardless of emotional state. Each action taken within a system compounds, creating steady movement toward goals. Momentum grows as repeated application builds both results and habit. Smart students learn that consistent structure beats sporadic brilliance every time.

Designing effective systems also enhances accountability and measurement. By clearly defining processes, milestones, and metrics, students can track progress and identify gaps. Each completed step provides evidence of forward motion, reinforcing competence and motivation. Momentum is maintained as students see tangible results from systematic action. Over time, these systems evolve, becoming more efficient and tailored to individual strengths and goals. Smart students use systems not just to act, but to optimize learning, performance, and growth.

Another advantage of designing systems is that they create automatic habits that reduce friction in daily action. Students no longer need to expend excessive mental energy deciding what to do or when to do it. Each repeated behavior becomes easier and more natural over time, reinforcing progress and skill development. Momentum grows as actions become habitual and predictable, allowing focus to shift toward higher-level problem-solving and creativity. Over time, systems make consistency effortless, turning intelligence into reliable output. Smart students leverage these structures to sustain action without relying solely on willpower.

Systems also provide a framework for scaling effort and achieving larger goals. When tasks are organized into repeatable processes, complex projects become manageable. Each small, systematic action contributes to bigger outcomes over time. Momentum is maintained as students consistently move forward, even when projects seem overwhelming. Over time, systems allow intelligence to be applied strategically, creating exponential growth and results. Smart students transform structured routines into engines of achievement and progress.

Designing systems also enhances clarity and focus. Students can prioritize what matters most while eliminating distractions and unnecessary tasks. Each element of a system directs attention toward high-impact actions, reducing wasted effort. Momentum grows as energy is consistently applied to meaningful work rather than being scattered across low-priority activities. Over time, clarity from structured systems improves decision-making and efficiency. Smart students use systems to channel intelligence into purposeful, deliberate action.

Systems also improve accountability and self-monitoring. By defining routines, deadlines, and checkpoints, students can measure progress and identify areas needing adjustment. Each step taken within a system provides evidence of effort, skill, and movement toward goals. Momentum is maintained as students continually refine processes based on feedback and results. Over time, this reinforces competence, confidence, and sustainable growth. Smart students transform structured action into measurable, continuous progress.

Another benefit of well-designed systems is the reduction of procrastination. Students often delay action when tasks feel overwhelming, unclear, or unstructured. Systems break goals into manageable, repeatable steps, making it easier to start and maintain momentum. Each step completed within a system reinforces action and diminishes hesitation. Momentum grows as the habit of structured, consistent effort replaces avoidance and indecision. Over time, students develop a proactive mindset where forward movement becomes automatic rather than forced.

Systems also foster resilience during setbacks or challenges. When unexpected obstacles arise, a system provides a framework for adjusting action without losing progress. Each step, no matter how small, continues the momentum toward the larger goal. Momentum is maintained as systems allow students to adapt while preserving forward motion. Over time, structured approaches help smart students weather challenges and maintain consistent growth. By relying on systems, intelligence is converted into steady, reliable progress regardless of external circumstances.

Another advantage is improved efficiency and time management. Systems allow students to organize tasks, routines, and priorities in ways that minimize wasted effort. Each action is guided by an established process, reducing indecision and distractions. Momentum grows as structured approaches optimize energy and focus for high-impact work. Over time, efficiency increases as systems streamline repetitive tasks and free mental bandwidth for creativity and learning. Smart students leverage systems to maximize results without expending unnecessary effort.

Designing systems also encourages continuous learning and refinement. Students can evaluate the effectiveness of routines and processes, adjusting them to improve outcomes. Each adjustment improves performance and builds confidence in decision-making. Momentum is reinforced as systems evolve in response to experience and insight. Over time, students create dynamic frameworks that adapt to goals, strengths, and challenges. Smart students transform structured approaches into tools for sustained growth and learning.

Another benefit of well-designed systems is that they create predictability, which reduces stress and mental load. Students no longer need to constantly decide what to do or worry about missing steps. Each repeated action becomes part of a reliable routine, freeing mental energy for problem-solving and creativity. Momentum grows as consistent, predictable processes allow focus on execution rather than planning in the moment. Over time, students feel less overwhelmed and more in control of their progress. Smart students leverage predictable systems to convert effort into steady, reliable achievement.

Systems also encourage prioritization of high-impact actions. By structuring tasks and routines, students can focus on what truly moves them toward their goals rather than being distracted by minor or low-value activities. Each step taken within the system contributes meaningfully to progress. Momentum is maintained as repeated, high-priority actions compound over time into significant outcomes. Over time, structured prioritization enhances both efficiency and effectiveness. Smart students use systems to align intelligence, effort, and results toward meaningful objectives.

Another advantage is the ability to scale effort and progress over time. As systems become established, students can layer additional tasks, projects, or challenges without losing consistency. Each small addition builds on the existing framework, expanding productivity and capability. Momentum grows as structured action compounds, creating exponential growth potential. Over time, smart students transform simple routines into powerful engines of learning, growth, and achievement. Systems become scalable tools that multiply the impact of intelligence and effort.

Designing systems also strengthens self-discipline and follow-through. By relying on routines and structures, students reduce the reliance on fluctuating motivation or willpower. Each completed step within a system reinforces the habit of action and accountability. Momentum is maintained as habit and structure replace hesitation and distraction. Over time, smart students cultivate disciplined, consistent action that supports long-term growth. Systems transform intelligence into practical, sustainable, and repeatable performance.

Another key advantage of systems is that they provide clarity during periods of uncertainty. Students often get stuck when they don't know where to start or what to do next. A well-designed system outlines the next actionable step, removing ambiguity and reducing stress. Each completed step reinforces direction and builds confidence. Momentum grows as students continue moving forward even when circumstances are unpredictable. Over time, systems transform intelligence into structured, dependable progress that doesn't rely on perfect conditions.

Systems also facilitate reflection and continuous improvement. By tracking progress and outcomes, students can identify what works and what doesn't within their routines. Each analysis informs adjustments that increase efficiency and effectiveness. Momentum is maintained as students iteratively refine their systems to achieve better results. Over time, this creates a culture of deliberate learning and growth. Smart students turn structured processes into adaptive, evolving frameworks that maximize achievement.

Another advantage is that systems reduce reliance on motivation and discipline alone. Students often struggle when they must rely solely on willpower to complete tasks. Systems embed action into routines, making progress automatic and less dependent on fluctuating mental states. Each step completed within a system reinforces the habit of consistent action. Momentum grows as structured routines carry students forward regardless of energy or motivation levels. Over time, intelligence is converted into consistent output through the power of systems.

Designing systems also encourages balance and sustainability. By planning routines and workflows, students can allocate energy effectively across work, rest, and personal growth. Each system that includes recovery and reflection prevents burnout and maintains long-term momentum. Momentum is reinforced as sustainable routines allow for continuous progress without exhaustion. Over time, students achieve more by combining structure with strategic pacing. Smart students leverage systems to create consistent, enduring, and balanced success.

Finally, building effective systems requires patience, experimentation, and commitment. Students must implement routines, observe results, and refine processes until they operate smoothly. Each improvement strengthens momentum, competence, and reliability. Momentum is sustained as evolving systems support action, learning, and progress. By designing systems that carry them forward, smart students transform intelligence into structured, measurable, and sustainable achievement, ensuring long-term growth and success.

QUOTES:

“Motivation is what gets you started. Habit is what keeps you going.”
— Jim Ryun

CHAPTER .26

THE POWER OF FINISHING SMALL THINGS

Many smart students focus on big ideas or ambitious goals, believing that only monumental achievements matter. This mindset often leads to unfinished projects, frustration, and wasted potential. The power of finishing small things lies in the momentum, confidence, and learning that accumulate from completion. Each small task completed reinforces the habit of follow-through and demonstrates tangible progress. Momentum grows as repeated small wins build both competence and motivation. Smart students learn that finishing consistently, even on minor tasks, produces far more impact over time than sporadic bursts of ambition.

One reason small completions are so powerful is that they provide immediate evidence of progress. Students who struggle to see results from large, long-term goals often feel discouraged or stuck. Finishing even minor tasks produces a sense of accomplishment that validates effort and builds confidence. Each completed step reinforces the understanding that action leads to results. Momentum is maintained as small wins compound into a track record of achievement. Over time, these incremental successes create a foundation for tackling larger goals with clarity and confidence.

Another consequence of ignoring small tasks is that they accumulate into mental clutter. Unfinished work creates stress, distraction, and decision fatigue. Completing small tasks clears space, reduces anxiety, and increases focus. Each finished item contributes to a sense of control and order, reinforcing motivation to continue. Momentum grows as students move from one small completion to the next, creating a chain of action that propels them forward. Smart students discover that clearing the small things first often makes room for larger breakthroughs.

Finishing small things also strengthens discipline and follow-through. Students who build the habit of completing tasks, no matter how minor, develop reliability and persistence. Each completed action reinforces the habit of seeing things through, even when motivation is low or obstacles appear. Momentum is maintained as the pattern of completion becomes automatic, making it easier to tackle progressively larger challenges. Over time, disciplined follow-through creates confidence and self-trust. Smart students turn consistent action into a dependable engine for long-term growth.

Another advantage of finishing small things is that it builds confidence through repeated proof of competence. Students who focus only on large, ambitious goals may feel inadequate if progress seems slow or uncertain. Completing smaller tasks provides concrete evidence that they can follow through and achieve results. Each small win reinforces self-belief and reduces fear of failure. Momentum grows as repeated completions demonstrate capability and reliability. Over time, this builds a resilient foundation of confidence that supports tackling larger, more complex challenges.

Finishing small tasks also strengthens focus and clarity. When students break larger projects into actionable pieces, they know exactly what needs to be done next. Each completed piece reduces overwhelm and directs attention toward the remaining steps. Momentum is maintained as progress becomes visible and measurable, guiding students toward completion. Over time, clarity from consistent small wins helps smart students manage even ambitious goals without feeling paralyzed. Systems of small completions create structure, focus, and continuous forward motion.

Another benefit is the creation of a feedback loop for learning and improvement. Each completed task provides information about what works, what doesn't, and how to improve. Students can analyze results, refine approaches, and apply lessons to future tasks. Momentum grows as repeated action produces both progress and insight. Over time, small completions accumulate into a rich body of experience, shaping smarter decisions and more effective strategies. Smart students leverage each finish, no matter how minor, to accelerate learning and growth.

Finishing small things also reduces procrastination and mental clutter. Unfinished tasks weigh on the mind, creating stress and distraction. Completing small items clears mental space and provides a sense of accomplishment that motivates continued action. Each task completed contributes to momentum, making it easier to tackle subsequent challenges. Over time, consistent small completions build habits of reliability, focus, and productivity. Smart students discover that small wins pave the way for bigger achievements and sustained progress.

Another key advantage of finishing small things is that it strengthens time management skills. Students learn to break larger tasks into manageable chunks, schedule them effectively, and allocate resources wisely. Each completed small task reinforces the ability to plan, prioritize, and execute efficiently. Momentum grows as students consistently move through their to-do list, building both skill and confidence. Over time, this approach reduces overwhelm and increases productivity. Smart students transform intelligence into consistent, well-managed action by mastering the art of small completions.

Finishing small tasks also improves motivation by creating a chain of visible accomplishments. Students often struggle to stay motivated when goals feel distant or abstract. Completing even minor tasks provides immediate feedback and a sense of achievement. Each small win fuels energy for the next step, reinforcing action and persistence. Momentum is maintained as repeated completions create a positive cycle of accomplishment and motivation. Over time, smart students develop the habit of action-driven motivation rather than relying solely on fleeting inspiration.

Another benefit is that small completions foster accountability. Students who focus on finishing tasks develop self-trust and reliability, as they can see tangible results of their efforts. Each completed task serves as evidence of their capability and follow-through. Momentum grows as students recognize that action produces measurable outcomes, reinforcing responsibility and discipline. Over time, this strengthens their ability to take ownership of projects and commitments. Smart students build a reputation of competence and dependability by consistently finishing what they start.

Finishing small things also encourages reflection and iterative improvement. Each completed task provides a chance to analyze performance, identify lessons, and adjust strategies for future action. Momentum is maintained as students apply insights from each small win to subsequent efforts, increasing efficiency and effectiveness. Over time, this continuous refinement develops skill, judgment, and expertise. Smart students leverage small completions not only for results but also as opportunities to improve processes, strategies, and outcomes.

Another advantage of finishing small things is that it helps overcome the fear of failure. Students often hesitate to start tasks because they worry about imperfection or mistakes. By focusing on completing small, manageable actions, the pressure of perfection diminishes. Each finished task demonstrates that progress is possible even if the outcome isn't flawless. Momentum grows as students realize that action, not perfection, drives results. Over time, smart students develop courage and resilience, turning fear into motivation through consistent completion.

Finishing small tasks also reinforces the habit of consistency. Students who regularly complete minor actions develop discipline and reliability that carry over to larger, more complex projects. Each small win strengthens the routine of follow-through, making it easier to maintain steady progress. Momentum is maintained as repetition compounds effort into measurable results. Over time, consistent action becomes second nature, reducing procrastination and boosting productivity. Smart students leverage consistent completion as a cornerstone for sustained achievement.

Another benefit is that small completions enhance focus and attention to detail. Students who aim to finish tasks learn to concentrate on the necessary steps and execute them thoroughly. Each completed action reinforces precision and care, improving the quality of work over time. Momentum grows as students develop both speed and accuracy through repeated practice. Over time, attention to detail in small tasks translates into competence in larger projects. Smart students turn careful, consistent completion into a foundation for excellence.

Finishing small things also strengthens problem-solving skills. Each completed task, no matter how minor, presents opportunities to encounter challenges, test solutions, and learn from results. Momentum is reinforced as students apply lessons learned to subsequent actions, improving efficiency and decision-making. Over time, repeated problem-solving experiences accumulate into greater competence and creativity. Smart students transform small actions into opportunities for skill growth and strategic thinking.

Another powerful outcome of finishing small things is the development of patience and long-term vision. Students who focus solely on big, immediate goals may feel frustrated by slow progress or delayed results. Completing small tasks teaches the value of incremental progress and builds understanding that large achievements are the sum of many small actions. Each finished task reinforces the principle that steady effort compounds into meaningful outcomes. Momentum grows as students internalize the habit of consistent follow-through, even when results aren't immediate. Over time, smart students cultivate the patience, discipline, and foresight necessary for sustained success.

Finishing small tasks also helps reduce overwhelm in complex projects. Large goals can feel paralyzing, causing students to procrastinate or overthink. By breaking tasks into manageable, finishable components, students create a clear path forward. Each small completion reduces mental clutter and builds clarity about next steps. Momentum is maintained as students move steadily through each component of a larger project. Over time, this approach makes ambitious goals achievable and less intimidating.

Another benefit is the reinforcement of accountability and self-trust. Students who consistently finish tasks, no matter the size, prove to themselves that they can follow through. Each small completion builds evidence of reliability and competence. Momentum grows as repeated success fosters confidence and reinforces the habit of action. Over time, students develop a reputation—and a personal standard—of dependability and discipline. Smart students learn that trust in themselves is built one finished task at a time.

Finishing small things also strengthens adaptability and problem-solving. Even minor tasks present challenges that require creative thinking, flexibility, and adjustment. Each completed task provides lessons about strategy, efficiency, and execution. Momentum is reinforced as students apply these lessons to future tasks, improving both speed and quality. Over time, these repeated learning experiences create greater resilience and capability. Smart students turn the act of finishing small things into a laboratory for continuous improvement and skill growth.

Finally, mastering the power of finishing small things sets the stage for larger successes. Repeated small wins build momentum, confidence, competence, and clarity, all of which support bigger goals. Each completed task is a building block that compounds over time into meaningful achievement. Momentum grows as the habit of completion becomes automatic and self-reinforcing. Over time, smart students transform intelligence into actionable results and continuous growth by embracing the practice of finishing consistently.

QUOTES:

"Little by little, one travels far." — J.R.R. Tolkien

CHAPTER .27

LEARNING HOW TO FAIL PRODUCTIVELY

Many smart students fear failure because they tie their identity to success and competence. This fear can lead to avoidance, perfectionism, and stagnation, preventing them from taking meaningful risks. Learning to fail productively means reframing failure as information, feedback, and opportunity for growth. Each failure, when approached intentionally, provides lessons that accelerate learning and improve future decisions. Momentum grows as students transform setbacks into stepping stones rather than barriers. Smart students understand that failure is not the opposite of success but a critical part of achieving it.

One reason productive failure is so important is that it accelerates learning. When students avoid mistakes, they also avoid opportunities to test their understanding, skills, and strategies. Each failed attempt reveals gaps in knowledge, areas for improvement, and potential alternative approaches. Momentum is maintained as students apply these lessons immediately to refine actions and strategies. Over time, students develop a deeper understanding of their field, skill, or craft. Smart students use failure as a tool to turn intelligence into practical, applied competence.

Another consequence of avoiding failure is the development of risk aversion. Students who fear making mistakes often play it safe, limiting opportunities for growth and innovation. By learning to fail productively, they embrace experimentation and calculated risk-taking. Each risk taken, even if it results in failure, provides valuable feedback and experience. Momentum grows as students learn to navigate uncertainty and make informed adjustments. Over time, this approach fosters courage, resilience, and adaptability. Smart students transform fear into action, using failure as a guide rather than a deterrent.

Failing productively also strengthens problem-solving and critical thinking. Each setback challenges students to analyze what went wrong, why it happened, and how to improve. Each reflection generates new strategies, insights, and approaches for future attempts. Momentum is reinforced as each failure informs subsequent action, creating a continuous cycle of learning and improvement. Over time, students become more resourceful, creative, and effective in tackling challenges. Smart students recognize that productive failure is not a setback but a mechanism for accelerated growth.

Another benefit of learning to fail productively is that it cultivates resilience. Students who encounter setbacks without fear are better able to recover and continue forward. Each failure provides an opportunity to practice bouncing back, adjusting strategies, and persevering. Momentum grows as students realize that temporary setbacks do not define their trajectory. Over time, resilience becomes a natural habit, enabling them to face increasingly complex challenges with confidence. Smart students transform obstacles into opportunities to strengthen both skill and character.

Failing productively also encourages experimentation and innovation. Students who are not afraid of failure are more willing to try new approaches, test hypotheses, and explore unconventional solutions. Each attempt, successful or not, generates insights that refine understanding and enhance future performance. Momentum is maintained as the lessons learned from experimentation inform subsequent action. Over time, this iterative process accelerates growth, skill acquisition, and creativity. Smart students leverage failure as a powerful tool to innovate and improve continuously.

Another advantage is the reduction of fear and perfectionism. Students who tie their self-worth to flawless outcomes often hesitate to act, waiting for ideal conditions. Learning to fail productively shifts the focus from being perfect to learning and progressing. Each failure normalized as part of growth reduces anxiety and encourages action. Momentum grows as students act consistently, experiment, and iterate without being paralyzed by fear. Over time, this mindset creates confidence, adaptability, and a willingness to tackle difficult challenges. Smart students replace perfectionism with progress-oriented action.

Failing productively also strengthens reflective thinking. After each setback, students analyze what went wrong, why it happened, and how to improve next time. Each reflection transforms failure into insight, guiding future strategies and decisions. Momentum is reinforced as lessons from past experiences are applied immediately to subsequent attempts. Over time, reflective practice sharpens judgment, efficiency, and skill. Smart students turn each failure into a practical opportunity for learning and mastery.

Another benefit of failing productively is that it develops humility and self-awareness. Students who embrace mistakes recognize that no one is infallible, and that growth comes from learning through experience. Each failure provides insight into strengths, weaknesses, and areas needing improvement. Momentum grows as students use these insights to refine strategies and skills. Over time, humility and self-awareness foster better collaboration, decision-making, and adaptability. Smart students combine competence with self-knowledge, allowing failure to become a catalyst for growth rather than a threat to identity.

Failing productively also encourages persistence over time. Students who view setbacks as temporary and instructive are less likely to give up when challenges arise. Each failure reinforces the lesson that sustained effort, reflection, and adjustment lead to eventual success. Momentum is maintained as students continue to act despite obstacles, building both skill and resilience. Over time, persistence becomes a habit, allowing students to tackle increasingly complex and ambitious goals. Smart students transform repeated failures into a roadmap for achievement.

Another advantage is improved problem-solving under pressure. When students experience failure, they must analyze situations, identify errors, and generate solutions quickly. Each failure challenges their critical thinking and forces creative approaches. Momentum grows as students apply these problem-solving skills to subsequent attempts, improving efficiency and outcomes. Over time, productive failure enhances adaptability and the ability to navigate uncertainty. Smart students use mistakes as opportunities to strengthen judgment, strategy, and execution.

Failing productively also strengthens emotional regulation. Students who fear failure often experience anxiety, frustration, or shame when mistakes occur. By reframing failure as a learning tool, emotional reactions shift from fear to curiosity and reflection. Each constructive approach to failure reinforces calmness, resilience, and confidence. Momentum is maintained as students respond to setbacks with clarity and composure rather than avoidance or panic. Over time, emotional regulation supports consistent action, problem-solving, and growth. Smart students integrate emotional control with practical learning to turn failure into progress.

Another benefit of failing productively is that it fosters a growth-oriented mindset. Students who view mistakes as learning opportunities begin to see intelligence and skill as expandable rather than fixed. Each failure becomes a source of knowledge that enhances ability and understanding. Momentum grows as students embrace challenges, experiment, and adapt based on real-world feedback. Over time, a growth mindset transforms setbacks into stepping stones rather than roadblocks. Smart students cultivate curiosity, resilience, and adaptability by reframing failure as essential to progress.

Failing productively also improves strategic planning. When students analyze why an attempt failed, they learn to anticipate obstacles, adjust methods, and allocate resources more effectively. Each reflection strengthens decision-making and problem-solving for future tasks. Momentum is maintained as students act with increasing insight and precision. Over time, this creates smarter, more efficient strategies that maximize results. Smart students use productive failure as a blueprint for refining approach and achieving greater success.

Another advantage is that it encourages experimentation without fear of judgment. Students often avoid trying new ideas because of the risk of embarrassment or criticism. By approaching failure productively, they become comfortable testing hypotheses and exploring alternatives. Each experiment, successful or not, generates learning and informs next steps. Momentum grows as repeated experimentation builds both confidence and competence. Over time, students develop creativity, innovation, and a willingness to take calculated risks. Smart students transform failure into a safe and powerful tool for discovery.

Failing productively also strengthens perseverance and grit. Students who repeatedly encounter setbacks and respond constructively learn that effort, reflection, and adjustment lead to long-term success. Each challenge overcome builds resilience and reinforces the habit of sustained action. Momentum is maintained as persistence becomes a natural response to difficulty rather than avoidance. Over time, grit ensures that smart students can pursue ambitious goals with determination, even in the face of repeated obstacles. By embracing productive failure, they convert challenges into lessons and action.

Another powerful outcome of learning to fail productively is the development of self-trust. Students who see themselves navigate setbacks successfully gain confidence in their judgment and abilities. Each failure handled constructively reinforces the understanding that they can overcome challenges and adapt effectively. Momentum grows as self-trust strengthens, allowing students to take risks without being paralyzed by fear. Over time, this internal confidence reduces reliance on external validation or reassurance. Smart students transform failure into a consistent source of empowerment, building the resilience and courage needed for long-term achievement.

Failing productively also encourages reflection on process rather than just outcomes. Students learn to examine not only what went wrong but why it happened and how the approach can be improved. Each analysis informs future strategies, making subsequent attempts more effective. Momentum is maintained as lessons from failure are applied immediately, creating a cycle of continuous improvement. Over time, students become adept at diagnosing challenges, adjusting methods, and optimizing performance. Smart students leverage the feedback inherent in failure to refine both thinking and action.

Another advantage is that it fosters adaptability in dynamic situations. Students who fear failure often stick rigidly to a plan, even when circumstances change. By embracing productive failure, they learn to adjust strategies in real time, respond to unexpected obstacles, and remain flexible. Each failure provides information that guides adaptation and decision-making. Momentum grows as students become comfortable navigating uncertainty and complexity. Over time, adaptability becomes a habit, allowing smart students to succeed even in unpredictable environments.

Failing productively also strengthens perseverance under pressure. Students who understand that failure is temporary and informative are less likely to give up when outcomes are unfavorable. Each setback becomes an opportunity to practice persistence, problem-solving, and determination. Momentum is maintained as challenges are approached with sustained effort and strategic adjustment. Over time, students cultivate grit and endurance, key qualities for achieving ambitious goals. Smart students convert repeated failures into stepping stones toward long-term success.

Finally, learning to fail productively requires consistent practice, reflection, and mindset development. Students must approach mistakes with curiosity, analyze outcomes, and apply lessons to future action. Each failure navigated constructively builds competence, confidence, and momentum. Momentum is sustained as setbacks are reframed as learning opportunities rather than threats. By mastering productive failure, smart students transform obstacles into tools for continuous growth, turning intelligence into applied skill, resilience, and enduring achievement.

Quotes:

"Only those who dare to fail greatly can ever achieve greatly." — Robert F. Kennedy

CHAPTER .28

USING CONSTRAINTS TO INCREASE CREATIVITY

Many smart students believe creativity requires unlimited freedom and endless options. Paradoxically, constraints often stimulate creativity more than complete freedom. Boundaries challenge the mind to find novel solutions within limits, forcing innovation and strategic thinking. Each constraint, whether a time limit, budget, or specific rule, provides a framework that channels energy into focused problem-solving. Momentum grows as students learn to leverage limitations instead of viewing them as obstacles. Smart students transform constraints into catalysts for original ideas, action, and progress.

One reason constraints enhance creativity is that they reduce overwhelm. Unlimited possibilities can paralyze decision-making and slow execution. By imposing boundaries, students focus on what truly matters and explore unconventional solutions within a defined space. Each constraint forces resourcefulness, clarity, and prioritization. Momentum is maintained as limitations guide action and reduce indecision. Over time, constraints become tools for structured experimentation and inventive thinking.

Constraints also encourage efficiency and strategic thinking. Students must find the most effective methods to achieve goals within limits, optimizing both resources and time. Each limitation requires careful planning, analysis, and decision-making. Momentum grows as students develop smarter approaches that maximize impact with minimal waste. Over time, this strengthens problem-solving skills, adaptability, and ingenuity. Smart students convert restrictions into opportunities for clever, resourceful solutions.

Another benefit is that constraints foster collaboration and perspective-taking. When boundaries are in place, students often need to seek input, brainstorm, and combine strengths to find solutions. Each challenge encourages teamwork, communication, and shared creativity. Momentum is reinforced as collective problem-solving produces ideas that would not emerge in isolation. Over time, students become more adept at leveraging diverse perspectives within defined limits. Smart students use constraints to enhance both creativity and cooperative skill.

Constraints also spark inventive thinking by forcing students to reconsider assumptions. When the usual methods or unlimited options are unavailable, the mind searches for alternative paths. Each imposed limitation encourages experimentation, problem-solving, and fresh perspectives. Momentum grows as students discover solutions they might have overlooked without boundaries. Over time, these exercises in creativity become habitual, sharpening adaptability and resourcefulness. Smart students use constraints to train their minds to innovate consistently under pressure.

Using constraints also enhances focus and efficiency. Students no longer waste time exploring irrelevant options or pursuing unnecessary avenues. Each restriction narrows attention to what matters most, enabling deeper engagement with the task at hand. Momentum is maintained as focused action produces tangible results faster and with greater clarity. Over time, the ability to work creatively within limits becomes a competitive advantage in problem-solving and innovation. Smart students convert limits into tools that enhance both productivity and originality.

Constraints also encourage experimentation with new methods and approaches. Students are prompted to test ideas they might otherwise ignore, because traditional solutions may not fit within the imposed boundaries. Each experiment, successful or not, produces learning, insight, and refined techniques. Momentum grows as students iterate and improve, using constraints to structure trial and error effectively. Over time, experimentation within limits fosters both skill development and creative confidence. Smart students transform restrictions into stepping stones for innovation and mastery.

Another benefit is that constraints build problem-solving resilience. Students learn to persist even when options are limited, developing flexibility and adaptability. Each obstacle forces critical thinking, evaluation of alternatives, and strategic decision-making. Momentum is reinforced as students overcome challenges systematically, turning limitations into opportunities. Over time, resilience and ingenuity become habitual, enabling students to handle increasingly complex problems. Smart students thrive in situations that demand both creativity and disciplined problem-solving.

Constraints also help students develop clarity in decision-making. When options are limited, they are forced to prioritize, weigh trade-offs, and choose the most effective course of action. Each decision made within constraints sharpens judgment and reinforces strategic thinking. Momentum grows as students learn to make faster, more confident choices under defined conditions. Over time, these habits enhance both efficiency and creativity in tackling complex problems. Smart students leverage constraints to simplify decisions while maximizing impact.

Using constraints also encourages focus on value rather than volume. Students learn to invest time and energy in the most meaningful or high-impact tasks. Each boundary clarifies what is essential, reducing wasted effort on less important actions. Momentum is maintained as concentrated effort produces measurable results more consistently. Over time, this approach cultivates skill, discipline, and efficiency. Smart students transform limitations into a framework for prioritizing and achieving what truly matters.

Another advantage is that constraints foster resilience against unexpected challenges. When students practice creativity within limits, they become comfortable adapting to unforeseen obstacles or restrictions. Each problem encountered within a constrained environment provides practice in flexibility and resourcefulness. Momentum grows as students apply these adaptive skills to new situations, maintaining forward movement despite setbacks. Over time, resilience becomes second nature, enabling sustained progress under varying conditions. Smart students turn limitations into opportunities to strengthen both creativity and problem-solving.

Constraints also support the development of innovative thinking under pressure. Students who work with boundaries must generate ideas quickly and efficiently, often arriving at solutions that are both creative and practical. Each challenge tackled within limits expands their repertoire of approaches and strategies. Momentum is reinforced as repeated practice builds confidence in the ability to innovate under constraints. Over time, smart students internalize structured creativity as a natural problem-solving skill. By embracing boundaries, they cultivate both originality and strategic insight.

Another benefit of using constraints is that they encourage resourcefulness and ingenuity. Students must find ways to achieve goals with limited tools, time, or information. Each challenge requires creative problem-solving and adaptation to the conditions at hand. Momentum grows as students learn to optimize what they have rather than rely on abundance. Over time, resourcefulness becomes a natural skill, allowing students to tackle complex problems with efficiency and creativity. Smart students leverage constraints to discover solutions that are clever, practical, and effective.

Constraints also promote experimentation with unconventional approaches. When standard methods don't fit within limits, students are pushed to think outside the box. Each experiment, whether successful or not, produces insight and refines strategies. Momentum is maintained as repeated experimentation generates both progress and learning. Over time, this habit fosters innovative thinking, flexibility, and confidence in tackling unfamiliar challenges. Smart students use boundaries as a catalyst to explore, test, and improve continuously.

Another advantage is that constraints enhance focus on outcomes rather than processes. Students must concentrate on achieving results within limits rather than getting lost in excessive planning or analysis. Each boundary clarifies the goal and streamlines decision-making. Momentum grows as students act efficiently toward meaningful objectives, producing tangible outcomes faster. Over time, this outcome-oriented mindset strengthens discipline, execution, and productivity. Smart students transform constraints into a structure that drives effective, results-focused action.

Constraints also encourage collaborative problem-solving. In group settings, limits often require teamwork, discussion, and combining different strengths to find solutions. Each challenge tackled collectively builds communication skills, mutual understanding, and creativity. Momentum is reinforced as collaborative efforts lead to results that might not have been achieved individually. Over time, students become skilled at leveraging diverse perspectives to innovate within boundaries. Smart students use constraints to strengthen both individual and group problem-solving abilities.

Another key advantage of using constraints is that they build confidence in problem-solving. When students consistently find solutions within limits, they develop trust in their ability to handle challenges. Each success reinforces the understanding that obstacles are manageable with creativity and strategy. Momentum grows as repeated experiences with constraints strengthen both competence and resilience. Over time, this confidence becomes internalized, allowing students to approach even unfamiliar situations with assurance. Smart students turn constraints into opportunities to prove to themselves that they can thrive under pressure.

Constraints also foster iterative learning and continuous improvement. Students learn to test ideas, assess results, and refine approaches within defined boundaries. Each attempt, whether it succeeds or fails, provides data to inform better strategies in the future. Momentum is maintained as each iteration builds skill, insight, and effectiveness. Over time, the habit of refining approaches within constraints develops adaptability, creativity, and mastery. Smart students leverage boundaries as a structured path for ongoing learning and innovation.

Another benefit is that constraints encourage prioritization and resource optimization. Students must evaluate what is most important and allocate limited resources effectively. Each decision made under constraints strengthens judgment and strategic thinking. Momentum grows as students learn to focus energy where it matters most, maximizing outcomes with minimal wasted effort. Over time, this skill becomes invaluable for managing complex projects, tight deadlines, and limited resources. Smart students transform limits into tools for disciplined, high-impact action.

Constraints also enhance creativity under pressure. When students face strict limitations, they must find novel ways to achieve goals, often producing more innovative solutions than they would with unlimited options. Each challenge sparks inventive thinking and problem-solving. Momentum is reinforced as repeated creative solutions build confidence, skill, and originality. Over time, students internalize the ability to innovate consistently, even in restrictive environments. Smart students use constraints as a framework to turn challenges into opportunities for creative excellence.

Finally, mastering constraints requires intentional practice, reflection, and experimentation. Students must observe results, refine methods, and strategically apply lessons learned to new situations. Each iteration strengthens competence, creativity, and momentum. Momentum is sustained as constraints guide structured learning and action toward tangible outcomes. Over time, smart students convert boundaries into powerful levers for growth, innovation, and sustained achievement. Constraints become not obstacles, but engines for progress, skill, and creative mastery.

QUOTES:

"Limitations live only in our minds. But if we use our imaginations, our possibilities become limitless." — Jamie Paolinetti

CHAPTER .29

HOW TO BUILD SKILLS OUTSIDE TRADITIONAL CLASSROOMS

Many smart students rely heavily on formal education, believing that classrooms and lectures are the only path to skill development. While traditional education provides knowledge, real competence is built through application, practice, and experimentation in the real world. Building skills outside the classroom allows students to explore, test, and apply knowledge in practical contexts. Each external experience creates feedback loops that accelerate learning and mastery. Momentum grows as students take ownership of their development and integrate theory with practice. Smart students recognize that skills acquired independently often become the most durable and valuable.

One reason learning outside traditional classrooms is powerful is that it encourages active engagement. Students are not passive recipients of information; they become creators, doers, and problem-solvers. Each task, project, or challenge requires deliberate action, critical thinking, and adaptation. Momentum is maintained as students apply knowledge in real situations, reinforcing understanding and competence. Over time, this experiential learning builds confidence, skill, and independence. Smart students transform intelligence into practical ability by seeking opportunities beyond structured lessons.

Another benefit is that external skill-building fosters creativity and innovation. Students who explore projects, hobbies, or side ventures are often exposed to diverse challenges that require original solutions. Each new experience forces creative thinking, experimentation, and adaptability. Momentum grows as students iterate and learn from both successes and failures. Over time, these experiences cultivate a mindset of exploration, problem-solving, and inventive action. Smart students leverage real-world experiences to expand capability, insight, and creative potential.

Building skills outside traditional classrooms also strengthens self-directed learning. Students must identify what to learn, how to practice, and when to seek feedback. Each decision enhances independence, initiative, and accountability. Momentum is maintained as students develop habits of curiosity, experimentation, and continuous improvement. Over time, self-directed learners become adaptable, resilient, and capable of acquiring new skills rapidly. Smart students transform intelligence into lifelong learning capacity through autonomous practice and reflection.

Another advantage of building skills outside traditional classrooms is the opportunity to learn through mistakes. In formal education, failure is often penalized or stigmatized, discouraging experimentation. Real-world practice allows students to test ideas, take risks, and learn from what doesn't work. Each mistake provides feedback that can be analyzed and applied to improve performance. Momentum grows as students use these lessons to refine strategies and increase competence. Over time, learning through trial and error outside the classroom develops resilience, problem-solving ability, and practical expertise.

Learning outside classrooms also encourages networking and collaboration. Students often engage with mentors, peers, and communities who share similar interests or expertise. Each interaction provides insights, feedback, and opportunities to practice skills in diverse contexts. Momentum is maintained as collaboration accelerates learning and exposes students to perspectives they might not encounter in formal settings. Over time, these relationships become invaluable for growth, opportunity, and skill development. Smart students leverage real-world connections to enhance both competence and confidence.

Another benefit is that external learning fosters adaptability and flexibility. The real world is rarely structured like a classroom, and students must adjust to changing circumstances, limited resources, and unexpected challenges. Each new scenario teaches problem-solving, strategic thinking, and creative adaptation. Momentum grows as students become comfortable navigating uncertainty and applying knowledge in dynamic contexts. Over time, adaptability becomes a habit, enabling students to handle complex problems with agility and skill. Smart students transform intelligence into practical capability through varied, unstructured experiences.

Building skills outside traditional classrooms also strengthens initiative and self-motivation. Students must identify what skills are valuable, how to acquire them, and when to practice or experiment. Each choice reinforces independence and the habit of taking action without external prompts. Momentum is maintained as students develop self-directed routines and pursue challenges proactively. Over time, initiative and self-motivation compound, creating a mindset oriented toward continual growth and achievement. Smart students turn curiosity and discipline into tangible skill development beyond structured learning environments.

Another benefit of building skills outside traditional classrooms is that it encourages exploration of diverse interests. Students are free to experiment with areas beyond standardized curricula, discovering passions and hidden talents. Each new experience provides insight into strengths, preferences, and potential career paths. Momentum grows as students integrate these explorations into their learning and personal growth. Over time, diverse experiences foster creativity, versatility, and broader problem-solving ability. Smart students leverage external learning to cultivate a multidimensional skill set that enhances both competence and fulfillment.

Learning outside formal education also promotes practical application of theoretical knowledge. Concepts learned in textbooks often remain abstract unless applied in real-world contexts. Each hands-on project, internship, or personal experiment transforms theory into tangible skill. Momentum is maintained as practical experience reinforces understanding, builds confidence, and encourages innovation. Over time, repeated application develops expertise that is both deep and usable. Smart students convert intelligence into actionable ability through consistent real-world practice.

Another advantage is the development of time-management and prioritization skills. When students pursue self-directed learning, they must balance practice, projects, and experimentation alongside other commitments. Each decision about how to allocate time and energy strengthens planning, focus, and discipline. Momentum grows as students create effective routines and habits that optimize learning and productivity. Over time, these skills become essential for managing complex tasks, long-term projects, and ambitious goals. Smart students transform self-directed learning into structured, efficient, and sustainable progress.

Building skills outside classrooms also fosters initiative and accountability. Students are responsible for setting goals, monitoring progress, and adjusting strategies without the structure of formal instruction. Each completed project or mastered skill reinforces self-trust, independence, and responsibility. Momentum is maintained as students consistently take ownership of their learning and outcomes. Over time, this habit of accountability becomes a cornerstone of personal and professional growth. Smart students convert autonomy into competence, confidence, and tangible achievement.

Another benefit of building skills outside traditional classrooms is the ability to customize learning to one's own pace and style. Students are no longer bound by the schedules, methods, or expectations of formal education. Each self-directed project allows for learning at an optimal speed and in ways that align with personal strengths. Momentum grows as students experiment, adjust, and refine their approach to maximize understanding and performance. Over time, this self-tailored learning enhances both mastery and confidence. Smart students leverage autonomy to turn intelligence into effective, personalized skill development.

Learning outside formal education also encourages curiosity-driven exploration. Students can follow questions, ideas, and interests that spark genuine engagement rather than adhering to prescribed curricula. Each inquiry-driven project strengthens critical thinking, problem-solving, and independent learning skills. Momentum is maintained as curiosity motivates sustained effort, experimentation, and application. Over time, this habit of following one's interests cultivates creativity, adaptability, and a passion for lifelong learning. Smart students transform natural curiosity into practical skills and meaningful achievements.

Another advantage is that external skill-building provides opportunities for real-world feedback. Unlike grades or exams, feedback from mentors, peers, clients, or the results of applied projects offers actionable insights. Each critique or outcome informs refinement and improvement, enhancing competence and effectiveness. Momentum grows as students use feedback to adjust strategies and build skill iteratively. Over time, this continuous feedback loop strengthens judgment, adaptability, and performance. Smart students convert external input into measurable progress and practical expertise.

Building skills outside classrooms also fosters problem-solving in complex, ambiguous environments. Students encounter challenges without predefined solutions, requiring creative thinking and initiative. Each problem tackled provides learning and insight that is difficult to replicate in structured classroom settings. Momentum is maintained as students gain confidence in navigating uncertainty and developing solutions independently. Over time, these experiences build resilience, critical thinking, and strategic skill. Smart students transform challenges outside traditional classrooms into opportunities for innovation and growth.

Another significant benefit of building skills outside traditional classrooms is the development of initiative and self-motivation. Students must take responsibility for identifying what to learn, seeking resources, and committing to consistent practice. Each completed project or learned skill reinforces the habit of taking action independently. Momentum grows as students develop confidence in their ability to create progress without external prompts. Over time, initiative and self-motivation compound, forming a foundation for lifelong learning and achievement. Smart students transform intelligence into proactive skill-building and tangible results.

Learning outside formal education also allows students to integrate interdisciplinary knowledge. Real-world challenges rarely fit neatly into a single subject, requiring students to combine insights from multiple fields. Each cross-disciplinary application strengthens problem-solving, creativity, and adaptability. Momentum is maintained as students discover connections, experiment, and apply integrated approaches to projects. Over time, this ability to combine knowledge across domains becomes a powerful advantage in both learning and career development. Smart students leverage external learning to build versatile, applicable expertise.

Another advantage is the opportunity to cultivate leadership and collaboration skills. Working on projects, internships, or community initiatives often requires coordinating with others, taking initiative, and managing responsibilities. Each experience reinforces communication, negotiation, and teamwork abilities. Momentum grows as students learn to influence, motivate, and contribute effectively in collaborative settings. Over time, these skills become essential for professional success and personal growth. Smart students turn real-world experiences into opportunities to develop both competence and leadership.

Building skills outside traditional classrooms also encourages risk-taking and resilience. Students are exposed to real-world uncertainty, requiring experimentation and tolerance for failure. Each attempt, whether successful or not, provides feedback and lessons that accelerate learning. Momentum is maintained as students refine strategies, overcome obstacles, and build confidence in handling challenges. Over time, resilience becomes habitual, allowing students to tackle increasingly ambitious goals without fear. Smart students convert risk and failure into practical growth and skill mastery.

Finally, learning outside classrooms fosters a mindset of continuous growth and curiosity. Students discover that skills, knowledge, and competence can always be expanded through action, experimentation, and reflection. Each new project, challenge, or experience reinforces the value of learning through doing. Momentum grows as students integrate knowledge, practice, and feedback into their personal and professional development. Over time, this mindset transforms intelligence into applied mastery, adaptability, and sustained achievement. Smart students recognize that building skills outside traditional classrooms is one of the most effective ways to turn potential into real-world results.

Chapter .30

Turning Curiosity into Output

Many smart students have endless curiosity but struggle to translate it into tangible results. Curiosity without action can become a cycle of endless learning, research, and thinking without producing meaningful outcomes. Turning curiosity into output means taking the questions, ideas, and insights that spark interest and channeling them into projects, products, or creations. Each small act of output reinforces understanding, skill, and momentum. Momentum grows as ideas are transformed into tangible results, providing evidence of progress and capability. Smart students recognize that intelligence multiplied by action produces meaningful results.

One reason converting curiosity into output is essential is that learning becomes deeper when applied. Simply consuming information or exploring topics mentally is only the first step; active application solidifies understanding. Each project, experiment, or creation requires synthesis, problem-solving, and decision-making. Momentum is maintained as students act on curiosity, transforming abstract knowledge into concrete skill. Over time, repeated action enhances both competence and confidence. Smart students turn questions into experiments and observations into achievements.

Turning curiosity into output also fosters creativity and innovation. Ideas that remain internal often stagnate or are forgotten, but expressing them externally invites refinement, testing, and improvement. Each attempt to create or produce, no matter how small, generates insight, feedback, and learning. Momentum grows as students iterate, adjust, and improve their outputs based on results. Over time, this practice cultivates originality, problem-solving ability, and inventive thinking. Smart students leverage curiosity as fuel for tangible creations, innovations, and solutions.

Another advantage is that producing output strengthens self-directed learning. Students decide what to explore, what questions to pursue, and how to act on discoveries. Each output requires planning, execution, and reflection, reinforcing autonomy and accountability. Momentum is maintained as action becomes a habit, turning curiosity into consistent learning and skill-building. Over time, self-directed output enhances both mastery and independence. Smart students transform their natural curiosity into disciplined, applied action.

Turning curiosity into output also helps students overcome procrastination and overthinking. When ideas remain in the mind, they can feel overwhelming or abstract, leading to indecision. By producing tangible results, even small ones, students break the cycle of rumination and take actionable steps forward. Each completed output reinforces the habit of acting on ideas rather than endlessly planning. Momentum grows as repeated action builds confidence, skill, and tangible evidence of progress. Over time, this habit transforms curiosity into consistent productivity and measurable achievement.

Producing output also provides opportunities for feedback and improvement. Ideas tested in the real world reveal strengths, weaknesses, and areas for refinement that remain invisible in thought alone. Each experiment or project delivers insights that guide future actions and strategies. Momentum is maintained as students apply feedback to iterate and improve, creating a continuous cycle of learning and growth. Over time, external input strengthens judgment, skill, and adaptability. Smart students turn curiosity into actionable learning through deliberate practice and reflection.

Another advantage is that output fosters accountability and commitment. When students commit to creating something—a project, report, prototype, or piece of writing—they set expectations for themselves. Each act of production requires follow-through, planning, and perseverance. Momentum grows as completing outputs reinforces reliability and self-trust. Over time, accountability strengthens habits, discipline, and confidence in tackling increasingly complex challenges. Smart students transform curiosity into disciplined, actionable progress that accumulates over time.

Turning curiosity into output also enhances creativity and problem-solving. Students must take raw ideas, observations, or questions and convert them into tangible forms, requiring analysis, synthesis, and adaptation. Each creation, experiment, or application challenges the mind to innovate and find solutions. Momentum is maintained as students refine and expand their outputs through iteration. Over time, this practice strengthens critical thinking, inventive ability, and practical skill. Smart students channel curiosity into actionable creations that produce both learning and real-world impact.

Turning curiosity into output also strengthens confidence in one's abilities. When students see ideas manifest into tangible results, they gain proof that they can take initiative and succeed. Each successful experiment, project, or creation reinforces self-efficacy and the belief that action produces meaningful outcomes. Momentum grows as students accumulate evidence of their capability, which fuels further exploration and risk-taking. Over time, confidence compounds, encouraging bolder, more ambitious outputs. Smart students transform curiosity into action, and action into a track record of achievement.

Producing output also deepens understanding and mastery. Students who act on ideas must apply knowledge, connect concepts, and solve problems in context. Each output requires translating theory into practice, which solidifies comprehension far beyond passive study. Momentum is maintained as repeated application strengthens skills and reinforces retention. Over time, this experiential approach creates both depth and versatility of understanding. Smart students convert curiosity into concrete learning experiences that build enduring competence.

Another advantage is that output provides opportunities to create value for others. Students who turn curiosity into tangible products, solutions, or insights can share, teach, or impact others. Each completed output has the potential to influence peers, mentors, or wider communities. Momentum grows as the process of creating value motivates further action and contribution. Over time, this external impact reinforces purpose, engagement, and accountability. Smart students leverage curiosity to not only grow themselves but also contribute meaningfully to the world around them.

Turning curiosity into output also encourages iterative improvement. Students rarely produce perfect results on the first attempt, and each iteration offers a chance to refine, optimize, and learn. Each cycle of creation and feedback builds skill, insight, and efficiency. Momentum is maintained as outputs are improved over time, increasing both quality and understanding. Over time, this iterative approach strengthens problem-solving, resilience, and innovation. Smart students turn initial ideas into polished results through persistent experimentation and refinement.

Turning curiosity into output also fosters problem-solving under real-world conditions. Students encounter challenges, constraints, and unexpected variables when applying ideas practically. Each problem solved strengthens critical thinking, creativity, and adaptive skills. Momentum grows as students learn to navigate complexity and uncertainty through action. Over time, this experience develops resilience, strategic thinking, and confidence in tackling unfamiliar situations. Smart students convert curiosity into practical problem-solving ability by producing actionable results.

Producing output also reinforces accountability to oneself and others. When students commit to turning ideas into tangible work, they create expectations for follow-through. Each completed project, experiment, or creation builds a habit of responsibility and reliability. Momentum is maintained as consistent action demonstrates competence and reinforces self-trust. Over time, this accountability becomes a cornerstone of both personal growth and professional effectiveness. Smart students turn curiosity into output that reflects both capability and integrity.

Another advantage is that output encourages exploration of new skills and methods. Students must experiment with tools, techniques, or approaches to bring their ideas to life. Each attempt expands knowledge, practical skill, and versatility. Momentum grows as experimentation leads to both innovation and mastery over new competencies. Over time, this continuous skill expansion makes students more adaptable, capable, and creative. Smart students leverage curiosity to explore, learn, and apply new abilities in real-world contexts.

Turning curiosity into output also strengthens reflection and learning from experience. Each act of creation provides feedback on what worked, what didn't, and what can be improved. Momentum is maintained as students analyze results, extract lessons, and apply insights to future projects. Over time, reflective practice deepens understanding, enhances skill, and increases the effectiveness of output. Smart students transform curiosity into actionable learning cycles that produce continuous improvement and measurable achievement.

Turning curiosity into output also cultivates perseverance and resilience. Students often encounter obstacles, failed experiments, or incomplete results when acting on ideas. Each challenge provides an opportunity to learn, adapt, and persist. Momentum grows as students develop the habit of continuing despite setbacks, treating obstacles as part of the creative process. Over time, perseverance becomes an internalized skill, allowing them to tackle increasingly complex projects and goals. Smart students transform curiosity into actionable resilience, ensuring that challenges fuel growth rather than create stagnation.

Producing output also strengthens the connection between learning and tangible achievement. Ideas, insights, and discoveries gain real value only when applied, tested, and shared. Each project or experiment demonstrates competence and provides measurable evidence of skill development. Momentum is maintained as students translate curiosity into visible, impactful results. Over time, this practice builds a track record of accomplishment that reflects both intelligence and applied ability. Smart students turn questions and ideas into concrete evidence of growth and capability.

Another advantage is that output encourages iterative improvement and mastery. Students rarely achieve perfection on the first attempt, and each iteration provides valuable feedback. Each refinement enhances skill, deepens understanding, and improves quality. Momentum grows as students learn from every attempt and continually adjust strategies for better outcomes. Over time, this iterative process develops expertise, creativity, and confidence in problem-solving. Smart students convert curiosity into a cycle of learning, experimentation, and increasingly polished results.

Turning curiosity into output also nurtures a sense of ownership and agency. Students are responsible for defining their goals, taking action, and creating results. Each completed project reinforces the belief that they can influence outcomes through effort and creativity. Momentum is maintained as students embrace ownership of their learning and its impact. Over time, this autonomy fosters confidence, independence, and self-directed growth. Smart students leverage curiosity to actively shape their skills, knowledge, and achievements.

Finally, mastering the process of turning curiosity into output transforms potential into real-world capability. Students learn to combine curiosity, intelligence, and action into meaningful results. Each act of creation builds skill, insight, and momentum for future endeavors. Momentum grows as students internalize the habit of transforming ideas into tangible outcomes. Over time, this practice develops competence, resilience, and creativity that extend far beyond formal education. Smart students understand that curiosity is powerful, but only when it is acted upon, producing tangible results, learning, and lasting achievement.

QUOTES:

"An idea not coupled with action will never get any bigger than the brain cell it occupied." — Arnold Glasow

CHAPTER .31

WHY FEEDBACK BEATS GRADES

Many smart students obsess over grades, believing they are the ultimate measure of intelligence and ability. While grades can provide a snapshot of performance, they rarely offer actionable guidance for improvement. Feedback, in contrast, provides detailed insight into strengths, weaknesses, and strategies for growth. Each constructive critique becomes a roadmap for skill development, helping students refine knowledge and execution. Momentum grows as students act on feedback, applying lessons to improve outcomes. Smart students understand that grades measure results, but feedback cultivates competence, confidence, and continuous progress.

One reason feedback is more valuable than grades is that it focuses on learning rather than judgment. Grades often reduce complex work to a single number or letter, masking nuances of understanding and effort. Feedback highlights specific areas for improvement, clarifies misunderstandings, and provides guidance for next steps. Momentum is maintained as students translate feedback into practical action, correcting mistakes and building new skills. Over time, this approach fosters mastery and deeper understanding rather than superficial achievement. Smart students prioritize learning through feedback rather than chasing numeric validation.

Feedback also encourages growth and experimentation. Students who fixate on grades may avoid risk, choosing safe or familiar approaches to ensure high marks. Constructive feedback, however, allows them to take chances, try new methods, and explore creativity without fear of permanent failure. Each experiment guided by feedback provides valuable insight and builds skill. Momentum grows as students iterate, learn, and refine strategies based on real input. Over time, this cycle of feedback and action strengthens problem-solving, innovation, and competence. Smart students leverage feedback as a tool for continuous improvement rather than a measure of worth.

Another advantage is that feedback develops self-awareness and metacognition. By reflecting on feedback, students learn to recognize patterns in their strengths, weaknesses, and learning habits. Each piece of input provides an opportunity to adjust strategies and approaches. Momentum is maintained as students incorporate insights from feedback into future work. Over time, this practice cultivates independent critical thinking and self-directed growth. Smart students transform external feedback into internalized understanding, enhancing both learning and decision-making.

Feedback also provides clarity on what matters most in learning and performance. Grades can be vague or overly generalized, offering little insight into specific areas of strength or weakness. Constructive feedback identifies exactly what to focus on, where to improve, and which strategies are effective. Each piece of guidance helps students prioritize effort and take targeted action. Momentum grows as students act on this clarity, refining skills and reinforcing progress. Over time, feedback turns vague effort into precise, impactful learning, making intelligence actionable.

Using feedback also strengthens the habit of reflection. Students must analyze comments, results, and observations to understand what changes will produce better outcomes. Each reflection deepens understanding, sharpens problem-solving, and highlights patterns in performance. Momentum is maintained as students integrate lessons from feedback into ongoing work. Over time, this reflective practice builds metacognitive skills, enabling smarter, faster learning. Smart students transform external feedback into internal insight that guides consistent improvement.

Another advantage is that feedback encourages experimentation and risk-taking. When students receive constructive guidance, they understand that mistakes are part of the learning process rather than permanent failures. Each experiment informed by feedback provides valuable data and enhances skill. Momentum grows as students iterate, test new approaches, and expand their competence. Over time, this cycle fosters creativity, resilience, and adaptive problem-solving. Smart students leverage feedback to explore possibilities safely and strategically, turning intelligence into practical capability.

Feedback also fosters accountability and ownership of learning. Students who actively seek and apply feedback demonstrate initiative and responsibility for their growth. Each application of guidance reinforces discipline, follow-through, and proactive learning. Momentum is maintained as students internalize the habit of acting on feedback rather than passively receiving grades. Over time, this accountability strengthens independence, confidence, and self-directed development. Smart students convert feedback into a system of continuous progress and personal mastery.

Feedback also cultivates resilience and emotional intelligence. Students who learn to accept constructive criticism develop the ability to manage disappointment and adapt to challenges. Each piece of feedback, whether positive or critical, becomes a tool for growth rather than a source of discouragement. Momentum grows as students practice responding to input with reflection and action instead of defensiveness. Over time, this builds emotional maturity, self-awareness, and persistence. Smart students transform external guidance into internal strength, enabling them to navigate setbacks and grow continuously.

Another advantage is that feedback encourages continuous improvement. Grades often signal an endpoint, creating the illusion of completion, while feedback identifies ongoing opportunities to refine skills and knowledge. Each constructive comment highlights areas for adjustment and experimentation. Momentum is maintained as students act on feedback iteratively, improving performance and deepening understanding. Over time, this habit produces expertise, adaptability, and measurable growth. Smart students convert intelligence into sustained progress by using feedback as a roadmap for mastery.

Feedback also strengthens collaboration and communication skills. When students seek, interpret, and discuss feedback with peers, teachers, or mentors, they learn to engage in meaningful dialogue about performance and development. Each interaction builds understanding, perspective, and the ability to give and receive input constructively. Momentum grows as students apply collaborative feedback to improve outcomes and refine ideas. Over time, these skills enhance both individual achievement and the capacity to work effectively with others. Smart students use feedback as a bridge to knowledge, skill, and stronger interpersonal capability.

Additionally, feedback helps students focus on learning rather than comparison. Grades often create competition and comparison with peers, which can limit motivation or lead to self-doubt. Constructive feedback centers attention on personal growth, skill development, and actionable steps. Each application of feedback redirects energy toward improvement rather than external validation. Momentum grows as students measure success by progress, competence, and results rather than ranking. Over time, this mindset fosters intrinsic motivation, self-directed learning, and confidence. Smart students prioritize feedback over grades, turning insights into meaningful development.

Feedback also encourages strategic thinking and problem-solving. When students receive detailed input, they learn to analyze challenges, identify patterns, and plan actionable steps for improvement. Each piece of feedback provides an opportunity to refine strategies and optimize performance. Momentum grows as students integrate insights into their decision-making and approach to tasks. Over time, this builds both analytical skill and adaptive intelligence. Smart students leverage feedback to turn curiosity, knowledge, and effort into effective solutions and measurable results.

Using feedback also develops persistence and long-term focus. Unlike grades, which are often a final judgment, feedback highlights areas for continuous refinement and improvement. Each iteration based on guidance encourages sustained effort, experimentation, and resilience. Momentum is maintained as students act repeatedly on constructive input, gradually mastering complex skills. Over time, this practice cultivates endurance, discipline, and the capacity to achieve ambitious goals. Smart students transform external evaluation into a roadmap for continuous growth.

Another advantage is that feedback fosters humility and openness to learning. Students learn to recognize that intelligence alone is not sufficient for mastery and that external perspectives provide valuable insight. Each piece of constructive criticism teaches students to accept limitations, consider alternatives, and expand their understanding. Momentum grows as students apply lessons learned without ego interference. Over time, humility and openness become habits that enhance collaboration, adaptability, and competence. Smart students use feedback to refine not only skills but also perspective and judgment.

Feedback also encourages accountability and ownership of progress. When students actively seek and respond to input, they take responsibility for their learning and outcomes. Each application of feedback reinforces the habit of self-directed action and disciplined follow-through. Momentum is maintained as students track improvements, measure results, and adjust strategies continuously. Over time, this builds reliability, independence, and confidence in managing personal growth. Smart students convert feedback into a system of self-improvement that drives sustained achievement.

Feedback also enhances adaptability and the ability to learn from diverse perspectives. Students who embrace input from teachers, peers, mentors, or real-world outcomes learn to adjust their approaches effectively. Each piece of feedback provides insight into alternative strategies, methods, or ways of thinking. Momentum grows as students experiment, refine, and iterate based on this guidance. Over time, adaptability becomes second nature, allowing learners to thrive in changing and complex environments. Smart students turn external evaluation into internal flexibility, strengthening both skill and resilience.

Using feedback encourages continuous curiosity and exploration. Constructive critiques often highlight gaps in knowledge or opportunities to deepen understanding. Each insight sparks further investigation, prompting students to explore new areas, ask questions, and expand expertise. Momentum is maintained as learning becomes an ongoing cycle of feedback, application, and discovery. Over time, this habit cultivates lifelong learning and intellectual agility. Smart students convert feedback into fuel for curiosity-driven action that produces real results.

Another advantage is that feedback fosters accountability to oneself and others. Students who seek and apply feedback develop responsibility for their growth, recognizing that progress depends on consistent action. Each instance of incorporating feedback reinforces commitment, focus, and ownership of learning. Momentum grows as students track their improvement and celebrate tangible progress. Over time, accountability becomes a habit, supporting personal discipline and long-term achievement. Smart students transform feedback into a mechanism for reliable, self-directed growth.

Feedback also strengthens the alignment of effort with goals. Instead of working blindly or only to meet external standards, students learn which actions have the greatest impact. Each piece of guidance helps prioritize efforts, refine techniques, and focus energy where it matters most. Momentum is maintained as students direct their work toward meaningful outcomes that reflect both learning and capability. Over time, this focus produces efficient, high-quality results that accelerate skill development. Smart students leverage feedback to ensure that intelligence and effort translate into tangible progress.

Finally, mastering the use of feedback completes the loop from intelligence to achievement. Students combine reflection, action, iteration, and improvement to transform knowledge into competence and mastery. Each application of guidance builds confidence, skill, and forward momentum. Momentum grows as students internalize feedback-driven habits, enabling continuous growth and measurable results. Over time, this practice produces adaptive, capable, and resilient learners prepared for complex challenges. Smart students understand that the true power of feedback lies not in evaluation itself, but in the action and growth it inspires.

QUOTES:

"Criticism, like rain, should be gentle enough to nourish a man's growth without destroying his roots."

— Frank A. Clark

PART V: BUILDING A FUTURE THAT FITS YOU

Chapter .32

Creating a Personal Definition of Success

Many smart students measure success by external standards: grades, accolades, or comparisons to peers. While these metrics provide temporary validation, they rarely capture personal growth, fulfillment, or purpose. Creating a personal definition of success means identifying what truly matters to you and aligning effort, goals, and actions with that vision. Each decision made with this clarity reinforces focus, motivation, and momentum. Momentum grows as students act intentionally rather than chasing arbitrary markers. Smart students understand that authentic success is defined by meaningful progress, not external approval.

One reason a personal definition of success is powerful is that it cultivates intrinsic motivation. When students pursue goals aligned with their values, interests, and strengths, effort becomes energizing rather than draining. Each step forward is reinforced by personal meaning rather than external reward. Momentum is maintained as students invest time and energy in what genuinely matters. Over time, intrinsic motivation sustains consistent action, learning, and achievement. Smart students turn self-defined purpose into the driving force behind effort and progress.

Creating a personal definition of success also encourages focus and clarity. Students can filter distractions, competing pressures, or irrelevant expectations by referencing their own criteria. Each choice, project, or opportunity is evaluated against this personal framework, ensuring alignment with meaningful objectives.

Momentum grows as students concentrate on high-impact actions rather than spreading energy across unrelated pursuits. Over time, this focus builds competence, mastery, and measurable outcomes. Smart students transform intelligence and effort into results that reflect their personal vision.

Another advantage is that defining success personally fosters resilience and confidence. When setbacks occur, students who rely on external measures of success may feel defeated or inadequate. Those with a self-defined framework can assess challenges in context, adapt strategies, and continue forward without losing motivation. Each obstacle overcome reinforces belief in personal judgment and capacity. Momentum is maintained as resilience grows from aligning effort with intrinsic goals. Over time, this practice cultivates self-trust, perseverance, and long-term achievement. Smart students rely on internal definitions of success to navigate challenges and sustain progress.

Creating a personal definition of success also encourages balance and well-being. When students rely solely on external markers, they often sacrifice health, relationships, and fulfillment in pursuit of approval. By defining success for themselves, they can include goals that nurture physical, emotional, and social well-being. Each decision made with this balance in mind strengthens sustainable momentum and long-term growth. Over time, students cultivate a life that integrates achievement with fulfillment. Smart students recognize that true success encompasses more than accomplishments—it includes holistic well-being.

Defining success personally also strengthens decision-making and prioritization. Students who know their values and goals can evaluate opportunities and challenges more effectively. Each choice is filtered through a framework that aligns with personal meaning, reducing wasted effort and regret. Momentum grows as students consistently act in ways that advance their authentic objectives. Over time, this practice cultivates clarity, efficiency, and focus in all aspects of life. Smart students convert self-awareness into purposeful action and meaningful results.

Another advantage is that personal definitions of success encourage autonomy and independence. Students are less reliant on validation from teachers, peers, or societal expectations. Each action taken toward a self-defined goal reinforces self-trust and accountability. Momentum is maintained as students develop confidence in their judgment and the ability to chart their own path. Over time, this autonomy fosters resilience, initiative, and lifelong learning. Smart students turn intelligence and curiosity into progress guided by their own vision rather than external pressure.

Creating a personal definition of success also nurtures motivation during setbacks and failures. External measures often feel arbitrary or discouraging when challenges arise, but intrinsic definitions provide context and perspective. Each setback becomes a learning opportunity rather than a judgment of worth. Momentum grows as students reflect, adapt, and continue pursuing goals aligned with their values. Over time, this approach fosters persistence, adaptability, and sustained achievement. Smart students use their personal definition of success as a compass through challenges, keeping momentum steady.

Defining success personally also empowers students to embrace individuality and authenticity. When goals are aligned with personal values and strengths, students feel free to pursue paths that resonate with them, rather than conforming to societal expectations. Each choice made with authenticity reinforces confidence and satisfaction. Momentum grows as students develop a sense of ownership over their journey, embracing their unique talents and perspectives. Over time, this fosters self-assurance, resilience, and creativity in decision-making. Smart students turn personal definitions of success into a foundation for authentic, meaningful achievement.

Creating a personal definition of success also enhances focus on progress rather than perfection. Students often become paralyzed by the fear of not meeting external standards or achieving flawless results. By measuring success according to personal growth and meaningful milestones, each step forward becomes significant, regardless of perfection. Momentum is maintained as students celebrate progress, learn from missteps, and continue acting toward goals. Over time, this mindset reduces stress, increases confidence, and strengthens persistence. Smart students prioritize progress over approval, transforming action into sustained development.

Another advantage is that personal definitions of success encourage intentional learning and skill-building. Students identify which abilities, knowledge, and habits directly contribute to their goals and invest time accordingly. Each focused effort produces tangible improvement aligned with their objectives. Momentum grows as students act deliberately, refining skills that matter most to their vision of success. Over time, this intentional approach builds competence, mastery, and confidence. Smart students transform intelligence into purposeful growth by aligning learning with personal goals.

Defining success personally also fosters resilience in the face of societal pressures and external comparisons. Students who rely on intrinsic measures of achievement are less swayed by rankings, judgments, or expectations from peers, family, or institutions. Each challenge becomes an opportunity to learn, adapt, and reaffirm personal priorities. Momentum is maintained as students remain committed to their own path despite external noise. Over time, this resilience cultivates independence, confidence, and sustained progress. Smart students leverage self-defined success to navigate pressures without losing direction or motivation.

Creating a personal definition of success also encourages reflection and self-awareness. Students must examine their values, strengths, interests, and long-term goals to determine what success truly means to them. Each reflection provides clarity, guiding decisions and prioritizing actions that align with personal meaning. Momentum grows as students continually reassess and adjust their objectives based on evolving insights. Over time, this habit fosters deeper understanding of oneself, purpose-driven learning, and intentional progress. Smart students use reflection as a compass to ensure their efforts consistently move them closer to authentic achievement.

Defining success personally also nurtures adaptability in an unpredictable world. As circumstances, opportunities, and challenges change, students with self-defined goals can adjust their path without losing sight of their purpose. Each adaptation reinforces flexibility, resilience, and strategic thinking. Momentum is maintained as students navigate change with confidence, using their personal framework as a guide. Over time, this adaptability strengthens problem-solving skills, creativity, and the ability to thrive in diverse environments. Smart students transform intelligence into action that evolves with life's demands.

Another advantage is that a personal definition of success strengthens intrinsic motivation for sustained effort. When students pursue goals that genuinely matter to them, energy and engagement are naturally maintained, even during setbacks or tedious work. Each task completed reinforces the meaningful connection between effort and purpose. Momentum grows as students stay committed to long-term objectives aligned with their values. Over time, this intrinsic motivation becomes a reliable engine for consistent learning, skill development, and achievement. Smart students turn personal meaning into a powerful driver of progress.

Creating a personal definition of success also fosters accountability and ownership. Students become responsible for defining their goals, tracking progress, and adjusting strategies as needed. Each action taken with intentional alignment strengthens habits of self-discipline and follow-through. Momentum is maintained as students measure success against their own standards rather than external expectations. Over time, this accountability cultivates independence, confidence, and mastery over personal and professional growth. Smart students convert personal clarity into consistent, results-driven action.

Creating a personal definition of success also empowers students to overcome fear of judgment and comparison. When success is measured by personal values and meaningful goals, external opinions and societal pressures lose their power. Each decision made in alignment with intrinsic standards reinforces confidence and self-trust. Momentum grows as students act boldly, pursuing objectives that resonate with their vision rather than conforming to expectations. Over time, this approach fosters independence, courage, and authentic achievement. Smart students transform external pressures into motivation for purposeful action instead of distraction or doubt.

Defining success personally also cultivates long-term vision and planning. Students learn to think beyond immediate rewards, grades, or recognition, focusing on meaningful milestones and enduring growth. Each goal is considered in the context of broader purpose, creating a roadmap for action and decision-making. Momentum is maintained as students connect daily effort to long-term objectives, ensuring consistent progress. Over time, this strategic perspective builds resilience, foresight, and the ability to achieve ambitious outcomes. Smart students turn intelligence and effort into sustainable success through purposeful planning and alignment.

Another advantage is that a personal definition of success fosters alignment between work and passion. When students identify what truly excites and motivates them, learning and effort become energizing rather than draining. Each project or challenge tackled with enthusiasm reinforces engagement, creativity, and mastery. Momentum grows as students pursue objectives that resonate deeply, producing both skill and satisfaction. Over time, this alignment cultivates sustainable growth, fulfillment, and resilience. Smart students transform curiosity, talent, and effort into meaningful accomplishments by following their passions.

Creating a personal definition of success also strengthens perseverance through challenges. Students guided by personal meaning can navigate setbacks and obstacles without losing direction or motivation. Each challenge becomes an opportunity to learn, adapt, and reaffirm commitment to intrinsic goals. Momentum is maintained as students respond proactively to difficulties, converting adversity into growth. Over time, perseverance and resilience compound, enhancing skill, confidence, and capacity for achievement. Smart students use personal clarity to remain steady and purposeful, even in the face of adversity.

Finally, defining success personally consolidates all aspects of growth—intellectual, emotional, and practical. Students align intelligence, curiosity, action, and reflection with meaningful goals. Each deliberate step contributes to mastery, fulfillment, and measurable results. Momentum grows as students integrate effort, skill-building, and purpose-driven action into their daily practice. Over time, this holistic approach produces adaptive, capable, and fulfilled individuals prepared for complex challenges and continuous growth. Smart students recognize that creating a personal definition of success transforms potential into enduring achievement and a purposeful life.

QUOTES:

"Success is liking yourself, liking what you do, and liking how you do it."

— Maya Angelou

CHAPTER .33

ESCAPING THE "SMART STUDENT" LABEL

Many students find themselves trapped by the label of "the smart one," which can create both pressure and limitations. While intelligence is often celebrated, it can also become a cage, forcing students to conform to expectations and avoid mistakes. Escaping this label means redefining identity, allowing growth, exploration, and learning without fear of judgment. Each choice to act beyond the constraints of the label reinforces autonomy and self-trust. Momentum grows as students recognize that intelligence alone does not define their potential or limit their abilities. Smart students learn that liberation comes when identity is flexible and not bound to external perceptions.

One reason escaping the "smart student" label is important is that it encourages risk-taking. When students feel obligated to maintain a perfect image, they may avoid challenges that could expose weaknesses. Letting go of the label allows experimentation, failure, and learning in areas outside their comfort zone. Each small risk taken reinforces courage, growth, and skill development. Momentum is maintained as students push boundaries without fear of judgment. Over time, this approach cultivates resilience, creativity, and practical competence. Smart students turn identity flexibility into a gateway for authentic progress.

Escaping the label also reduces the fear of failure and imperfection. Students often tie their self-worth to being seen as intelligent, making mistakes feel like personal shortcomings. By separating identity from performance, mistakes become opportunities for learning rather than threats to self-image. Each misstep embraced as part of growth reinforces confidence, adaptability, and curiosity. Momentum grows as students develop comfort with imperfection and experimentation. Over time, this mindset strengthens resilience, independence, and skill mastery. Smart students understand that failure is a tool for growth, not a reflection of worth.

Another advantage is that shedding the label fosters self-directed learning. Students can pursue interests, skills, and challenges based on curiosity and purpose, rather than external expectations of what a "smart student" should do. Each choice made independently reinforces autonomy, initiative, and motivation. Momentum is maintained as students explore new areas, tackle diverse projects, and expand capability. Over time, self-directed action builds competence, creativity, and confidence. Smart students convert freedom from labels into practical growth and meaningful achievement.

Escaping the "smart student" label also allows students to embrace curiosity over performance. When intelligence is tied to identity, students may focus on appearing knowledgeable rather than asking questions or exploring new ideas. Letting go of the label encourages experimentation, inquiry, and learning without fear of judgment. Each act of curiosity reinforces confidence, skill development, and momentum. Over time, this mindset cultivates creativity, adaptability, and a love for learning. Smart students learn that true growth comes from exploration, not merely maintaining an image of intelligence.

Another advantage is that shedding the label reduces pressure and stress. The expectation to always perform perfectly or excel academically can create anxiety and burnout. By separating identity from perceived intelligence, students can approach challenges with less fear and more enjoyment. Each step taken without pressure strengthens resilience, focus, and consistent effort. Momentum grows as students engage fully in learning, experimenting, and creating without self-imposed limitations. Over time, this practice produces both competence and well-being. Smart students convert freedom from expectation into sustainable progress and confidence.

Escaping the "smart student" label also enhances risk tolerance in learning and life. Students freed from the constraints of perfection are more willing to try new skills, tackle unfamiliar subjects, and pursue unconventional paths. Each successful experiment reinforces courage and capability, while each failure becomes a source of insight rather than shame. Momentum is maintained as students build skill through exploration and applied effort. Over time, this mindset fosters resilience, innovation, and adaptability. Smart students transform liberated thinking into action, learning, and meaningful results.

Letting go of the label also strengthens self-awareness and identity flexibility. Students learn to see themselves as multi-dimensional, capable of growth beyond innate intelligence or early achievements. Each new endeavor expands identity, skill, and confidence. Momentum grows as students embrace challenges without being confined by others' perceptions. Over time, this flexibility fosters creativity, self-trust, and the ability to adapt to changing circumstances. Smart students understand that intelligence is only one part of their potential, not the definition of who they are.

Escaping the "smart student" label also encourages collaboration and openness to learning from others. When students feel pressured to always appear knowledgeable, they may avoid asking questions or seeking help. Letting go of the label allows them to engage openly, share ideas, and learn from peers and mentors. Each collaborative experience strengthens understanding, communication, and teamwork skills. Momentum grows as students integrate diverse perspectives into their own learning. Over time, this practice enhances both competence and relationships. Smart students turn identity freedom into a foundation for richer learning and stronger connections.

Another benefit is that shedding the label fosters resilience to criticism and setbacks. Students who tie their identity to being smart often take criticism personally, leading to discouragement or avoidance. By separating self-worth from perceived intelligence, they can interpret feedback constructively. Each challenge or critique becomes an opportunity to grow rather than a threat to identity. Momentum is maintained as students respond to obstacles with learning and adaptation. Over time, resilience becomes habitual, enabling sustained growth and achievement. Smart students convert external evaluation into actionable insight rather than fear.

Escaping the label also allows for greater experimentation and creativity. When intelligence is a fixed identity, students may stick to familiar areas where they feel competent. Letting go encourages exploration, risk-taking, and creative problem-solving in new domains. Each attempt, whether successful or not, builds experience, confidence, and skill. Momentum grows as students pursue innovative projects without fear of judgment. Over time, this mindset fosters original thinking, adaptability, and practical achievement. Smart students transform freedom from labels into inventive and meaningful action.

Letting go of the "smart student" label also encourages balanced self-perception. Students learn to value effort, growth, and curiosity alongside intelligence. Each recognition of progress and learning reinforces confidence and motivation. Momentum is maintained as students measure achievement by personal growth and applied learning rather than external validation. Over time, this approach cultivates humility, self-awareness, and sustained development. Smart students turn balanced self-perception into a framework for continuous improvement and fulfillment.

Escaping the "smart student" label also encourages self-compassion and reduces fear of imperfection. Students often tie mistakes or struggles to their identity as intelligent, creating pressure to appear flawless. By letting go of the label, they can treat errors as learning opportunities rather than personal failures. Each instance of self-compassion strengthens resilience, confidence, and willingness to take risks. Momentum grows as students experiment, fail, and try again without fear or shame. Over time, this practice fosters both emotional well-being and sustainable growth. Smart students turn self-compassion into a tool for consistent learning and achievement.

Another benefit is that shedding the label fosters curiosity-driven learning. Freed from the pressure to always perform or impress, students can explore topics, skills, and ideas for the sake of understanding and enjoyment. Each exploration enhances knowledge, creativity, and practical ability. Momentum is maintained as students pursue curiosity without fear of judgment or expectation. Over time, this habit develops intellectual flexibility, problem-solving skills, and the ability to adapt to new challenges. Smart students convert freedom from labels into a lifelong passion for learning and growth.

Escaping the "smart student" label also strengthens authentic confidence. Confidence rooted in identity rather than intelligence is fragile and dependent on external validation. By redefining self-worth around effort, learning, and action, students cultivate a stable, internalized sense of capability. Each challenge faced and skill developed reinforces authentic self-confidence. Momentum grows as students approach tasks with assurance and curiosity rather than fear of judgment. Over time, authentic confidence supports resilience, initiative, and meaningful achievement. Smart students transform their identity from a static label into a dynamic source of empowerment.

Letting go of the label also encourages adaptability in goals and pursuits. Students are free to pivot, change focus, and take on new challenges without feeling constrained by the expectation of maintaining a "smart student" persona. Each adjustment reinforces flexibility, problem-solving, and resourcefulness. Momentum is maintained as students respond to changing circumstances with creativity and strategic thinking. Over time, adaptability becomes a habit, allowing students to thrive in unpredictable environments. Smart students use identity flexibility to navigate life's complexities while continuing to grow and succeed.

Escaping the "smart student" label ultimately allows students to define their own path and purpose. Freed from expectations tied to intelligence, they can pursue goals that align with values, curiosity, and long-term growth. Each decision made independently reinforces autonomy, motivation, and momentum. Momentum grows as students take intentional action toward meaningful objectives rather than trying to maintain an image. Over time, this approach cultivates resilience, creativity, and sustained achievement. Smart students understand that liberation from labels transforms intelligence from a fixed trait into a dynamic tool for growth and meaningful success.

Letting go of the label also strengthens intrinsic motivation over extrinsic validation. Students who no longer measure themselves solely by accolades or comparison develop the drive to learn, create, and improve for their own sake. Each action taken for intrinsic reasons reinforces engagement, persistence, and skill-building. Momentum is maintained as students focus on personal growth rather than external approval. Over time, intrinsic motivation becomes a reliable engine for consistent effort and achievement. Smart students convert freedom from expectation into purpose-driven action that produces lasting results.

Escaping the "smart student" label encourages courage to face new challenges. When intelligence is tied to identity, students may avoid unfamiliar or difficult tasks to protect their reputation. By freeing themselves from the label, they can embrace uncertainty and experimentation. Each challenge tackled builds experience, skill, and confidence. Momentum grows as students develop resilience and adaptability through repeated action. Over time, this courage fosters innovation, problem-solving, and meaningful accomplishment. Smart students turn liberated thinking into bold, purposeful growth.

Another advantage is that shedding the label promotes holistic learning and personal development. Students can cultivate creativity, emotional intelligence, practical skills, and interpersonal growth alongside cognitive ability. Each effort contributes to a more versatile, capable, and well-rounded individual. Momentum is maintained as students integrate multiple dimensions of growth into everyday practice. Over time, this holistic approach produces adaptive, competent, and fulfilled learners. Smart students leverage freedom from labels to maximize potential across all areas of life.

Finally, escaping the "smart student" label creates a sustainable foundation for long-term achievement. Students act with purpose, curiosity, and resilience rather than fear of judgment or expectation. Each step toward personal goals reinforces competence, confidence, and momentum. Over time, this practice produces adaptive, capable, and fulfilled individuals who can navigate complex challenges. Smart students understand that identity flexibility is the key to turning intelligence into continuous, meaningful growth and success.

Chapter .34

Choosing Direction Without Needing Certainty

Many smart students feel stuck because they wait for complete certainty before making decisions. The desire to choose the "perfect" path can lead to paralysis, overthinking, and missed opportunities. Choosing direction without needing certainty means acting with clarity about priorities while accepting that not every outcome is predictable. Each decision made in motion reinforces momentum, learning, and confidence. Momentum grows as students realize that progress comes from action, not from perfect foresight. Smart students understand that uncertainty is not a barrier but a natural part of growth and exploration.

One reason this approach is powerful is that it reduces fear of failure. Students often delay choices because they worry about making the wrong decision and damaging their future. By embracing action despite uncertainty, each step becomes an experiment rather than a high-stakes judgment. Momentum is maintained as students test assumptions, learn from outcomes, and adjust strategies in real time. Over time, this mindset cultivates resilience, adaptability, and practical problem-solving skills. Smart students turn uncertainty into opportunity by focusing on learning and movement rather than perfection.

Choosing direction without needing certainty also fosters clarity through action. Students gain insight about themselves, their interests, and their strengths by engaging in decisions and projects. Each choice produces information, feedback, and experience that refines future decisions. Momentum grows as students use evidence from action to guide progress, rather than waiting for external guarantees. Over time, this iterative approach strengthens judgment, confidence, and self-trust. Smart students understand that clarity emerges from doing, not from overanalyzing.

Another advantage is that this mindset encourages flexibility and adaptability. When students act without requiring certainty, they learn to pivot, iterate, and adjust as circumstances evolve. Each adjustment reinforces learning, problem-solving, and creativity. Momentum is maintained as students embrace change as part of growth rather than seeing it as a setback. Over time, adaptability becomes a habit, enabling students to navigate complex, unpredictable challenges successfully. Smart students convert uncertainty into a dynamic environment for skill-building and achievement.

Choosing direction without needing certainty also strengthens decision-making skills over time. Students who act despite ambiguity learn to weigh options, consider consequences, and make informed choices quickly. Each decision, even if imperfect, provides experience that refines judgment and intuition. Momentum grows as students gain confidence in navigating complexity and uncertainty. Over time, this practice builds decisiveness, strategic thinking, and self-trust. Smart students convert action in uncertainty into a skill set that supports both personal and professional growth.

Another advantage is that this mindset encourages learning from mistakes rather than fearing them. Waiting for certainty often prevents action, while acting without full knowledge inevitably produces errors and feedback. Each misstep becomes an opportunity to reflect, adapt, and improve. Momentum is maintained as students treat mistakes as experiments that provide valuable insight. Over time, this approach develops resilience, problem-solving ability, and adaptive expertise. Smart students understand that mistakes are not failures but stepping stones to growth.

Choosing direction without needing certainty also cultivates initiative and proactive behavior. Students learn to create opportunities and take action rather than waiting for ideal conditions or instructions. Each initiative reinforces confidence, competence, and the habit of movement. Momentum grows as students engage in projects, decisions, and experiments that produce real results. Over time, initiative becomes a default behavior, increasing both skill and achievement. Smart students transform uncertainty into motivation to act, rather than a reason to hesitate.

This mindset also strengthens emotional resilience. Students who act without full certainty develop comfort with discomfort, ambiguity, and risk. Each encounter with the unknown builds tolerance, patience, and adaptability. Momentum is maintained as students learn to manage anxiety and hesitation, focusing instead on process and learning. Over time, this resilience supports bold decision-making and sustained effort. Smart students convert uncertainty into an opportunity to grow mentally, emotionally, and practically.

Choosing direction without needing certainty also fosters creativity and innovation. When students are not bound by the need to have the "perfect" answer or path, they are free to explore unconventional approaches and ideas. Each experiment or decision taken under uncertainty encourages problem-solving and novel thinking. Momentum grows as students test possibilities, iterate, and refine strategies based on real-world results. Over time, this mindset cultivates original thought, adaptability, and practical ingenuity. Smart students transform the discomfort of uncertainty into opportunities for inventive solutions and meaningful progress.

Another advantage is that acting without full certainty builds confidence in personal judgment. Students learn that they can make decisions, assess outcomes, and adjust their approach without needing external validation. Each decision reinforces trust in their own reasoning and capacity to navigate ambiguity. Momentum is maintained as confidence grows, reducing hesitation and second-guessing. Over time, this approach strengthens self-reliance, autonomy, and proactive action. Smart students convert uncertainty into assurance that they can handle complex challenges successfully.

Choosing direction without needing certainty also encourages iterative learning. Students who take action quickly gain feedback and insights that would be impossible to acquire by waiting for perfect information. Each step produces data that informs the next, creating a cycle of continuous improvement. Momentum grows as students refine their approach through repeated action and reflection. Over time, iterative learning builds mastery, adaptability, and practical skill. Smart students leverage uncertainty as a feedback-rich environment to accelerate growth and competence.

This mindset also strengthens problem-solving under real-world conditions. Life rarely presents fully predictable scenarios, and students who act despite uncertainty practice navigating complexity and ambiguity. Each challenge tackled without complete certainty develops reasoning, strategy, and resilience. Momentum is maintained as students build confidence in managing uncertainty effectively. Over time, this approach fosters capability, resourcefulness, and the ability to thrive in dynamic situations. Smart students turn uncertainty into a training ground for practical problem-solving and decision-making.

Choosing direction without needing certainty also encourages a mindset of exploration and opportunity. Students who act without waiting for perfect conditions open themselves to new experiences, knowledge, and skills. Each choice taken under uncertainty provides insight that informs future decisions and personal growth. Momentum grows as students continuously experiment and adapt, turning the unknown into productive learning. Over time, this approach cultivates curiosity, adaptability, and practical intelligence. Smart students leverage uncertainty as a doorway to exploration and meaningful achievement.

Another benefit is that acting without certainty strengthens resilience and emotional stability. Students learn to tolerate ambiguity, setbacks, and unexpected outcomes without being paralyzed by fear or self-doubt. Each challenge faced reinforces the ability to respond effectively and maintain momentum. Momentum is maintained as students develop confidence in navigating unpredictability and managing emotions under pressure. Over time, this habit builds resilience, problem-solving, and self-trust. Smart students transform uncertainty into a source of strength rather than a barrier to action.

Choosing direction without needing certainty also promotes decisiveness and initiative. Students learn to act promptly, evaluate outcomes, and iterate rather than endlessly deliberating. Each decision, even imperfect, reinforces experience, judgment, and capacity for action. Momentum grows as students gain confidence in their ability to make effective choices in complex situations. Over time, decisiveness becomes a habitual skill, empowering students to seize opportunities and navigate challenges. Smart students convert uncertainty into proactive movement, driving both learning and achievement.

This mindset also enhances learning through real-world feedback. Students who act without full certainty gain firsthand experience, discover consequences, and identify effective strategies more quickly than those who wait for perfect conditions. Each action provides information, insight, and improvement opportunities. Momentum is maintained as students apply lessons learned to refine skills and approaches. Over time, this process strengthens practical competence, adaptability, and applied intelligence. Smart students treat uncertainty as a laboratory for experimentation, growth, and mastery.

Choosing direction without needing certainty ultimately reinforces a growth-oriented mindset. Students learn that progress is built through action, reflection, and adaptation, rather than waiting for ideal conditions or guarantees. Each decision made in motion contributes to skill development, experience, and self-confidence. Momentum grows as students see tangible results from acting despite ambiguity. Over time, this habit cultivates resilience, practical intelligence, and sustained achievement. Smart students understand that mastery and meaningful progress come from moving forward, not from waiting for certainty or perfection.

This mindset also encourages exploration of passions and interests. Freed from the pressure to find the one "perfect path," students can experiment with different areas of study, work, or creative pursuits. Each exploration provides new insight, skills, and clarity about what truly aligns with personal goals. Momentum is maintained as students pursue a variety of experiences that build competence and perspective. Over time, this openness fosters versatility, adaptability, and self-discovery. Smart students turn uncertainty into opportunity by exploring widely while refining focus over time.

Acting without full certainty also strengthens decision-making under pressure. Students learn to assess information quickly, weigh trade-offs, and make choices confidently even when outcomes are unknown. Each decision reinforces judgment, reasoning, and the ability to act decisively. Momentum grows as students gain evidence of their capacity to navigate complexity and ambiguity effectively. Over time, this skill becomes a core competence for both personal and professional success. Smart students convert uncertainty into a training ground for practical, high-stakes decision-making.

Choosing direction without needing certainty also cultivates independence and self-trust. Students rely on their own judgment, learning from outcomes and iterating as needed, rather than waiting for external validation. Each successful navigation of ambiguity reinforces autonomy and confidence in one's ability to manage challenges. Momentum is maintained as students trust their decisions and adjust strategies proactively. Over time, independence becomes a habit, supporting initiative, leadership, and lifelong learning. Smart students leverage uncertainty to build self-reliance and confidence in navigating complex paths.

Finally, embracing action without certainty produces long-term growth and sustainable momentum. Students develop resilience, adaptability, judgment, and practical intelligence through repeated engagement with uncertain situations. Each step forward compounds into skill, insight, and capability. Momentum grows as students learn to convert ambiguity into actionable opportunity and continuous improvement. Over time, this approach creates adaptive, confident, and accomplished individuals. Smart students understand that progress and success are generated by consistent action, learning, and iteration, not by waiting for perfect clarity or certainty.

Chapter .35

Designing a Life Around Strengths, Not Expectations

Many smart students live under the weight of others' expectations, trying to meet standards set by parents, teachers, or society. While these expectations may be well-intentioned, they often pull students away from their natural strengths and passions. Designing a life around personal strengths means identifying abilities, talents, and interests that energize and motivate action. Each choice aligned with strengths reinforces confidence, momentum, and authentic growth. Momentum grows as students focus energy on what they do best, producing meaningful results and satisfaction. Smart students learn that fulfillment comes from leveraging strengths, not conforming to external pressures.

One reason this approach is powerful is that it maximizes effectiveness and efficiency. When students act in alignment with their strengths, effort produces disproportionately greater results compared to forcing work in areas of weakness. Each task or project completed using natural abilities builds competence, confidence, and momentum. Momentum is maintained as students prioritize areas where they have the greatest impact. Over time, this practice produces mastery, achievement, and sustainable growth. Smart students convert self-knowledge into a roadmap for high-impact action and progress.

Designing a life around strengths also fosters motivation and engagement. Students are more likely to persist, take initiative, and enjoy work that aligns with their natural talents. Each activity that engages strengths reinforces intrinsic interest and energy. Momentum grows as students immerse themselves in meaningful work, developing skills while enjoying the process. Over time, this alignment produces sustained effort, deeper learning, and long-term accomplishment. Smart students turn strengths into a source of motivation, creativity, and consistent achievement.

Another advantage is that this approach reduces stress, burnout, and frustration. When students focus on expectations that do not match their abilities, they expend excessive energy with limited results. By centering life around strengths, effort becomes rewarding rather than draining. Each action taken in alignment with natural abilities reinforces confidence, satisfaction, and efficiency. Momentum is maintained as students experience success through work that resonates with their skills. Over time, this habit produces resilience, well-being, and sustainable achievement. Smart students transform alignment with strengths into a practical formula for fulfillment.

Designing a life around strengths also enhances creativity and problem-solving. When students operate in areas where they are naturally skilled, they can think more freely, innovate, and approach challenges from a position of competence. Each application of strength reinforces insight, experimentation, and practical solutions. Momentum grows as students leverage their abilities to tackle increasingly complex tasks with confidence. Over time, this practice produces originality, skill mastery, and meaningful impact. Smart students convert natural talent into actionable creativity that drives success.

Another advantage is that focusing on strengths builds resilience in the face of setbacks. When students encounter obstacles in areas aligned with their talents, they are more likely to persist, adapt, and find alternative strategies. Each challenge navigated through strength reinforces confidence and determination. Momentum is maintained as students continue forward, learning from difficulties rather than being discouraged by them. Over time, this habit produces enduring competence, problem-solving ability, and self-trust. Smart students transform natural ability into a platform for resilience and growth.

Designing a life around strengths also improves learning efficiency. Students absorb new information faster, retain skills more effectively, and develop mastery quicker when engaging areas of innate aptitude. Each new skill built on strengths compounds previous knowledge, reinforcing momentum. Momentum grows as students experience tangible progress and repeated success. Over time, this approach accelerates growth, performance, and achievement. Smart students use their strengths as a lever to enhance learning and create lasting competence.

This approach also encourages clarity and prioritization in life decisions. When students understand their strengths, they can evaluate opportunities, projects, and paths based on alignment with what they do best. Each choice made in alignment with strengths reduces wasted energy on pursuits that offer little impact. Momentum is maintained as students focus on high-value actions that produce meaningful outcomes. Over time, this focus develops discipline, strategic thinking, and purposeful achievement. Smart students convert self-awareness into intentional action that maximizes potential.

Designing a life around strengths also encourages strategic risk-taking. When students operate from a position of competence, they can take on challenges that stretch their abilities without being overwhelmed. Each calculated risk provides new experience, learning, and growth opportunities. Momentum grows as students gain confidence in their capacity to tackle bigger, more complex tasks. Over time, this mindset cultivates boldness, adaptability, and innovative problem-solving. Smart students turn the power of their strengths into leverage for meaningful progress and achievement.

Another advantage is that focusing on strengths strengthens self-confidence. Students who consistently apply their natural abilities see tangible results, which reinforces belief in their skills and judgment. Each accomplishment achieved through strengths validates effort and reinforces momentum. Momentum is maintained as students approach new challenges with assurance in their capability. Over time, this confidence compounds, supporting leadership, initiative, and resilience. Smart students transform recognition of their abilities into a reliable foundation for sustained growth and success.

Designing a life around strengths also fosters alignment between passion and action. When students identify areas where they excel and enjoy the work, effort becomes energizing rather than draining. Each task completed in alignment with passion and strength reinforces engagement and motivation. Momentum grows as students pursue meaningful goals that resonate with both their abilities and interests. Over time, this approach produces deep satisfaction, sustained effort, and authentic achievement. Smart students leverage this alignment to turn intelligence and talent into purposeful, fulfilling results.

This approach also encourages lifelong learning and development. Students who focus on strengths naturally seek to refine and expand these areas, building mastery over time. Each new skill, insight, or technique acquired compounds existing ability, reinforcing momentum. Momentum is maintained as students continue to develop expertise in areas that matter most to them. Over time, this habit produces a trajectory of growth, competence, and capability that extends across personal and professional life. Smart students use their strengths as a foundation for continuous learning and improvement.

Designing a life around strengths also reduces frustration and wasted effort. When students focus on tasks that don't align with their abilities, progress is slow and energy is drained. By identifying and leveraging strengths, each action produces greater results with less struggle. Momentum grows as students experience repeated success and reinforcement in areas where they excel. Over time, this approach builds confidence, efficiency, and motivation. Smart students convert knowledge of their abilities into consistent, high-impact action that drives growth.

Another benefit is that focusing on strengths encourages clarity in goal-setting. When students understand what they do best, they can set objectives that are realistic, meaningful, and aligned with natural abilities. Each decision guided by this understanding reinforces direction, purpose, and momentum. Momentum is maintained as students act consistently on goals that maximize their potential. Over time, this clarity develops strategic thinking, discipline, and effective planning. Smart students leverage their strengths to design a roadmap for sustained achievement and personal fulfillment.

Designing a life around strengths also nurtures intrinsic motivation and engagement. Students naturally enjoy and feel energized by work in areas where they are capable and passionate. Each activity reinforces interest, skill, and commitment. Momentum grows as students continue to invest effort in pursuits that feel meaningful and rewarding. Over time, this engagement produces mastery, creativity, and sustained performance. Smart students turn strengths into a source of internal drive that propels learning and achievement.

This approach also strengthens adaptability and growth potential. By focusing on strengths, students can identify complementary areas for development that enhance their natural abilities. Each new skill acquired builds on existing competencies, creating a compounding effect that reinforces momentum. Momentum is maintained as students continuously expand capacity while staying grounded in areas of strength. Over time, this strategy produces versatile, competent, and resilient learners. Smart students use their strengths as a base to grow in multiple dimensions of life and work.

Designing a life around strengths also fosters a sense of ownership and autonomy. When students identify and act on their unique abilities, they take responsibility for their growth and decisions. Each choice made in alignment with strengths reinforces independence, confidence, and momentum. Momentum is maintained as students rely on their own judgment and initiative rather than external validation or expectations. Over time, this practice cultivates self-trust, strategic thinking, and resilience. Smart students understand that leveraging their strengths empowers them to create a life guided by purpose, mastery, and authentic achievement.

Another advantage is that focusing on strengths reduces comparison and envy. Students who operate from their own abilities are less likely to measure themselves against others, as success is defined personally. Each action grounded in strengths reinforces self-worth and confidence. Momentum grows as students direct energy toward meaningful progress rather than external competition. Over time, this habit develops humility, clarity, and self-assuredness. Smart students convert understanding of strengths into a stable framework for personal and professional fulfillment.

Designing a life around strengths also nurtures creativity and innovation. Students operating in areas of talent are more confident to experiment, take risks, and explore novel solutions. Each creative attempt builds skill, insight, and momentum. Momentum is maintained as students expand their capabilities through experimentation and application. Over time, this mindset produces adaptive thinking, ingenuity, and effective problem-solving. Smart students leverage their strengths as a launchpad for original ideas and meaningful impact.

This approach also strengthens long-term resilience. Students who focus on strengths are better equipped to handle challenges because they can rely on proven abilities and adaptive strategies. Each difficulty navigated reinforces confidence, skill, and momentum. Momentum grows as students face setbacks without losing direction or motivation. Over time, resilience becomes a habitual trait, supporting sustainable achievement. Smart students transform awareness of strengths into stability and perseverance during uncertain or challenging situations.

Finally, designing a life around strengths promotes sustained engagement and personal satisfaction. Students who act in alignment with natural abilities find energy, focus, and fulfillment in their endeavors. Each achievement reinforces the connection between effort, talent, and meaning. Momentum grows as students continue to pursue work that resonates with their skills and passions. Over time, this alignment produces mastery, purpose, and enduring success. Smart students recognize that building life and work around strengths is the key to turning potential into consistent, meaningful, and impactful achievement.

QUOTES:

"Don't chase what you're told; chase what you're good at and passionate about."

— Anonymous

Chapter .36

Learning How to Start Before You're Ready

Many smart students delay action because they feel they are not fully prepared, skilled, or knowledgeable. Waiting for the "perfect moment" often results in procrastination, missed opportunities, and stagnation. Learning how to start before you're ready means embracing imperfection, taking small steps, and prioritizing action over preparation. Each step taken builds confidence, experience, and momentum. Momentum grows as students discover that progress comes from doing, not waiting for certainty. Smart students understand that readiness emerges through action, not prior planning or perfect preparation.

One reason this approach is essential is that it transforms fear into movement. Students often hesitate because they fear failure, judgment, or inadequacy. By starting anyway, even with limited resources or skills, they confront fear and transform it into learning. Each small action reinforces courage, self-trust, and skill acquisition. Momentum is maintained as students continue moving forward despite discomfort. Over time, this habit develops resilience, confidence, and practical capability. Smart students convert hesitation into action, accelerating growth and achievement.

Starting before you're ready also encourages iterative learning. Students learn faster when they engage with real-world tasks rather than waiting to master every detail beforehand. Each attempt produces feedback, insight, and opportunities for adjustment. Momentum grows as students refine skills, strategies, and understanding through experience. Over time, iterative learning produces mastery, competence, and adaptability. Smart students leverage action as a learning tool, gaining knowledge and skill faster than by waiting for perfect preparation.

Another advantage is that starting before readiness fosters creativity and problem-solving. When students dive into action without all the answers, they are forced to think creatively and find solutions in real time. Each challenge faced under incomplete preparation develops resourcefulness, ingenuity, and confidence. Momentum is maintained as students gain experience in navigating uncertainty and adapting strategies. Over time, this mindset produces innovation, resilience, and practical intelligence. Smart students transform imperfection into opportunity by learning through action.

Starting before you're ready also reduces procrastination and overthinking. When students wait to feel fully prepared, they often spend excessive time planning, researching, or worrying. By acting immediately, even imperfectly, they break the cycle of delay and begin generating results. Each step forward reinforces confidence, experience, and momentum. Momentum grows as students learn that progress depends on doing, not on waiting for ideal conditions. Over time, this habit develops decisiveness, initiative, and a practical sense of capability.

Another advantage is that this approach encourages experimentation and learning through trial and error. Students discover what works, what doesn't, and how to improve by engaging directly with tasks. Each attempt produces insight, feedback, and opportunities for adjustment. Momentum is maintained as students refine strategies and solutions based on real-world experience. Over time, iterative experimentation fosters adaptability, problem-solving, and innovation. Smart students convert action, even when imperfect, into a powerful tool for rapid growth.

Starting before readiness also cultivates resilience and comfort with discomfort. Students learn to tolerate uncertainty, mistakes, and challenges without being paralyzed by fear or perfectionism.

Each experience handled under less-than-ideal conditions strengthens patience, flexibility, and emotional fortitude. Momentum grows as students confront obstacles with proactive effort rather than hesitation. Over time, resilience becomes a habitual response, enabling sustained progress under complex circumstances. Smart students transform the discomfort of imperfection into a driving force for action and growth.

This approach also fosters self-trust and autonomy. Students realize they are capable of making decisions, navigating challenges, and adjusting strategies without needing external validation or perfect conditions. Each successful action, no matter how small, reinforces belief in their judgment and ability to handle uncertainty. Momentum is maintained as students gain confidence through repeated practice and experimentation. Over time, self-trust becomes a reliable foundation for independent learning, problem-solving, and achievement. Smart students leverage early action to strengthen competence and confidence simultaneously.

Starting before you're ready also enhances creativity and innovation. When students act without having every detail planned, they are forced to improvise and think outside conventional solutions. Each challenge navigated under incomplete preparation strengthens problem-solving skills and encourages experimentation. Momentum grows as students generate ideas, test them, and refine their approach in real time. Over time, this mindset fosters originality, adaptability, and practical intelligence. Smart students turn imperfect action into a laboratory for creativity and meaningful results.

Another benefit is that this approach encourages momentum through small wins. Students who begin despite uncertainty quickly discover that even minor progress builds confidence and motivation. Each small accomplishment reinforces the ability to act and improves skill through direct experience. Momentum is maintained as students continue to achieve incremental results that compound into significant progress. Over time, this practice produces a habit of consistent action and self-reinforcing growth.

Smart students leverage early movement to create ongoing energy and direction toward larger goals.

Starting before readiness also strengthens the ability to prioritize and focus. When students take action immediately, they learn to identify the most critical tasks and make effective decisions on the fly. Each choice reinforces clarity, resourcefulness, and practical judgment. Momentum grows as students continuously engage in high-impact activities rather than waiting for a perfect plan. Over time, this skill develops discipline, strategic thinking, and a results-oriented mindset. Smart students convert early action into a structured path for learning and achievement.

This mindset also reduces fear of failure and perfectionism over time. Students learn that progress is more valuable than flawless execution, and that mistakes are opportunities for feedback and improvement. Each step forward reinforces resilience and a proactive attitude toward challenges. Momentum is maintained as students accumulate experience, insight, and confidence through practical engagement. Over time, this habit fosters courage, adaptability, and self-efficacy. Smart students transform hesitation into action, turning fear into a tool for learning and growth.

Starting before you're ready also develops the habit of decisive action. Students learn to move forward even when information is incomplete or conditions are imperfect, reducing hesitation and overanalysis. Each decision taken reinforces confidence, initiative, and momentum. Momentum grows as students act consistently, learning to trust their judgment and refine strategies through experience. Over time, this practice builds decisiveness, autonomy, and practical intelligence. Smart students convert action under uncertainty into a reliable habit that drives progress and achievement.

Another advantage is that this approach encourages incremental learning and continuous improvement. Students gain feedback and insight from each step, allowing them to adapt and improve as they go. Each attempt, whether successful or flawed, contributes to skill development and understanding. Momentum is maintained as students accumulate experience and adjust strategies in real time. Over time, this iterative approach strengthens competence, creativity, and problem-solving. Smart students leverage imperfect starts to build a compounding cycle of learning and growth.

Starting before readiness also fosters courage and confidence in facing challenges. Students realize they can handle tasks and obstacles without complete preparation, which reduces fear and builds resilience. Each challenge confronted under uncertainty strengthens emotional and practical capability. Momentum grows as students continue to act despite discomfort or ambiguity. Over time, this mindset produces self-assured, adaptable learners who embrace complexity. Smart students transform initial hesitation into courage, consistently converting uncertainty into opportunity.

This mindset also enhances adaptability and flexibility. Students who act before they feel ready learn to pivot, iterate, and adjust strategies based on real-world outcomes. Each adjustment strengthens judgment, creativity, and problem-solving skills. Momentum is maintained as students navigate change without losing direction or energy. Over time, adaptability becomes a habitual skill that supports long-term learning, growth, and achievement. Smart students turn early action into a foundation for continuous adjustment and improvement.

Starting before you're ready ultimately transforms mindset, skill, and achievement. Students discover that action itself generates readiness, experience, and confidence, rather than waiting for perfect conditions. Each step forward builds competence, momentum, and self-trust. Momentum grows as students accumulate evidence of their capability and ability to adapt. Over time, this habit produces resilient, proactive, and capable individuals who thrive in uncertainty. Smart students understand that starting—even imperfectly—is the key to unlocking growth, learning, and real-world success.

Another benefit is that this approach fosters independence and initiative. Students learn to rely on themselves to navigate challenges, make decisions, and iterate strategies without waiting for guidance. Each independent action reinforces autonomy, self-confidence, and momentum. Momentum is maintained as students take ownership of progress and outcomes. Over time, independence becomes a habit, empowering students to pursue goals with clarity and determination. Smart students convert early action into a foundation for lifelong self-directed learning and achievement.

Starting before readiness also cultivates resilience in the face of setbacks. Students quickly learn that mistakes, challenges, and unexpected outcomes are part of the learning process. Each encounter with imperfection strengthens problem-solving, adaptability, and persistence. Momentum grows as students continue moving forward, refining strategies, and applying lessons learned. Over time, resilience becomes second nature, enabling students to navigate complex or unpredictable situations effectively. Smart students turn the inevitability of mistakes into a tool for sustained growth and competence.

This mindset also enhances creativity and resourcefulness. Students who begin without full preparation are forced to innovate, find solutions, and make decisions with limited information. Each situation tackled under uncertainty builds ingenuity, practical intelligence, and confidence. Momentum is maintained as students accumulate experiences that reinforce adaptive thinking. Over time, this approach produces highly creative, capable, and strategic learners. Smart students leverage imperfect starts to cultivate skills and insights that waiting could never provide.

Finally, learning to start before you're ready encourages continuous momentum and compounding growth. Each small action taken generates feedback, skill, and confidence that accelerate future efforts. Momentum grows as students act consistently, learning, iterating, and improving over time. Over the long term, this habit produces adaptive, resilient, and highly capable individuals. Smart students recognize that consistent action —regardless of initial readiness—is the engine that transforms potential into meaningful, lasting success.

CHAPTER .37

MANAGING ENERGY INSTEAD OF TIME

Many smart students focus on managing their schedules rather than their energy, assuming that more hours automatically translate into better results. However, productivity, learning, and creativity depend on how well energy is allocated, not just time spent. Managing energy instead of time means recognizing personal rhythms, prioritizing high-impact work during peak energy periods, and allowing for rest and recovery. Each choice made to optimize energy reinforces focus, efficiency, and momentum. Momentum grows as students learn to align effort with their natural capacity, rather than forcing output at suboptimal times. Smart students understand that sustainable performance comes from energy management, not simply clocking more hours.

One reason this approach is powerful is that it reduces burnout and fatigue. Students who ignore energy levels often overcommit, work long hours without focus, and struggle to sustain effort. By tracking energy and scheduling tasks accordingly, each action becomes more productive and less draining. Momentum is maintained as students leverage high-energy periods for critical work and allow recovery when needed. Over time, this habit cultivates stamina, focus, and resilience. Smart students convert awareness of energy into sustainable, high-quality performance.

Managing energy also enhances cognitive performance and creativity. Students are more likely to think deeply, solve problems, and generate original ideas when their energy levels are optimized. Each period of high energy supports sharper focus, better decision-making, and more effective learning. Momentum grows as students align their most demanding or creative work with peak energy moments. Over time, this practice produces more consistent performance and higher-quality outcomes. Smart students turn energy awareness into a tool for smarter, not harder, work.

Another advantage is that this approach fosters balance and well-being. Students who manage energy effectively incorporate rest, recovery, and self-care into their routines, reducing stress and improving health. Each period of intentional recovery restores focus, creativity, and motivation. Momentum is maintained as students prevent depletion, ensuring sustainable engagement in learning and work. Over time, this balance builds resilience, satisfaction, and long-term productivity. Smart students recognize that rest is not wasted time but an essential component of high performance.

Managing energy instead of time also improves focus and concentration. Students who work when their energy is high can immerse themselves fully in tasks, reducing distractions and improving output quality. Each session of focused effort produces better results in less time, reinforcing momentum. Momentum grows as students learn to identify their peak energy periods and schedule critical tasks accordingly. Over time, this practice strengthens discipline, attention, and effectiveness. Smart students understand that aligning work with energy maximizes both productivity and learning.

Another advantage is that energy management reduces mental fatigue and decision fatigue. When students track their energy, they avoid attempting demanding tasks during low-energy periods, preventing mistakes and frustration. Each strategic choice to work with energy, rather than against it, preserves cognitive resources and momentum. Momentum is maintained as students maintain clarity, creativity, and problem-solving ability throughout the day. Over time, this habit develops resilience, consistency, and mental endurance. Smart students convert awareness of energy into smarter, sustainable decision-making and performance.

Managing energy also fosters better emotional regulation and stress management. Students who respect their energy levels are less likely to experience frustration, irritability, or overwhelm. Each intentional break or rest period allows emotions to reset, maintaining motivation and momentum. Momentum grows as students respond to challenges calmly and effectively rather than reacting impulsively. Over time, emotional stability becomes a natural skill, supporting resilience and productive engagement. Smart students recognize that energy management is as much about mental and emotional well-being as it is about output.

This approach also enhances creativity and problem-solving ability. When energy is high, students are more likely to think flexibly, generate new ideas, and approach problems with innovative solutions. Each high-energy session becomes an opportunity to make meaningful progress and advance complex projects. Momentum is maintained as students consistently leverage energy for their most demanding cognitive tasks. Over time, this habit fosters ingenuity, insight, and high-level performance. Smart students turn energy awareness into a competitive advantage for creativity and critical thinking.

Managing energy instead of time also strengthens self-awareness and personal insight. Students learn to notice patterns in their focus, motivation, and stamina, which allows them to plan work more effectively. Each observation about energy levels informs decisions about when and how to tackle tasks. Momentum grows as students align their efforts with periods of high alertness and engagement. Over time, this practice develops intuition, self-discipline, and the ability to optimize performance naturally. Smart students turn awareness of personal energy rhythms into a strategic advantage for learning and achievement.

Another benefit is that this approach encourages intentional scheduling and planning. Students prioritize high-value tasks during peak energy periods and save low-energy moments for routine or restorative activities. Each decision to match task difficulty with energy availability reinforces momentum and effectiveness. Momentum is maintained as students maximize output and minimize wasted effort. Over time, intentional scheduling becomes a habit, increasing productivity and maintaining balance. Smart students understand that timing and energy alignment are more critical than simply filling hours with work.

Managing energy also reduces stress and enhances resilience. Students who respect their limits avoid overloading themselves, which minimizes frustration and mental exhaustion. Each strategic decision to rest or recover sustains focus, motivation, and emotional stability. Momentum grows as students maintain consistent engagement without hitting burnout. Over time, resilience and stress management become second nature, supporting sustained achievement. Smart students convert energy management into a tool for both performance and well-being.

This approach also improves decision-making and problem-solving quality. Students are sharper, faster, and more creative when their energy is optimized. Each task approached during high-energy periods produces better results and reinforces confidence. Momentum is maintained as students consistently make high-quality choices that advance goals efficiently. Over time, this habit strengthens judgment, adaptability, and overall effectiveness. Smart students recognize that peak performance relies on energy alignment, not just effort or intelligence.

Managing energy instead of time also fosters creativity in problem-solving and innovation. Students who work during high-energy periods can approach challenges with clarity, flexibility, and fresh perspective. Each period of focused energy allows for experimentation, idea generation, and strategic thinking. Momentum grows as students use their best energy for the tasks that require the most insight and problem-solving ability. Over time, this practice develops ingenuity, adaptability, and resourcefulness. Smart students leverage their energy rhythms to produce work that is both creative and impactful.

Another advantage is that energy management improves learning retention and comprehension. Students who engage in mentally demanding tasks when their energy is high are better able to absorb information and retain knowledge. Each learning session taken at peak energy strengthens memory, understanding, and application. Momentum is maintained as students make consistent progress without wasted effort. Over time, this approach develops cognitive efficiency and mastery. Smart students recognize that energy, not just time, drives effective learning and skill development.

Managing energy also strengthens emotional and mental resilience. Students learn to respect their limits and build recovery periods into their schedules, preventing frustration and burnout. Each intentional rest or low-energy task reinforces balance, focus, and motivation. Momentum grows as students maintain consistent effort without succumbing to exhaustion or stress. Over time, this habit produces emotional stability, patience, and long-term sustainability. Smart students convert energy-awareness into resilience, ensuring they can navigate challenges effectively over time.

This approach also supports strategic prioritization. Students who track energy can allocate their best effort to high-value tasks and use lower-energy periods for routine or supportive work. Each decision to align energy with task demands reinforces efficiency and effectiveness. Momentum is maintained as students focus on impactful work during peak capacity. Over time, this strategy strengthens judgment, discipline, and the ability to achieve meaningful results consistently. Smart students understand that energy alignment maximizes impact while minimizing wasted effort.

Managing energy instead of time ultimately strengthens overall life satisfaction and well-being. Students who align work with their energy levels experience less stress, greater focus, and more consistent performance. Each decision to prioritize energy over hours worked reinforces balance, resilience, and momentum. Momentum grows as students maintain consistent, high-quality effort without depleting themselves. Over time, this habit produces sustainable achievement, confidence, and fulfillment. Smart students recognize that energy-aware action, not sheer time investment, is the key to meaningful, lasting success.

Another benefit is that energy management encourages self-reflection and awareness. Students begin to notice when they are most alert, motivated, and effective, and they adjust routines to optimize these periods. Each insight into personal rhythms strengthens decision-making, prioritization, and momentum. Momentum is maintained as students apply these insights to both daily tasks and long-term projects. Over time, self-awareness becomes a habitual tool for maximizing effort and outcomes. Smart students turn understanding of their energy patterns into a competitive advantage for performance and learning.

Managing energy also reinforces proactive planning and intentionality. Students learn to schedule demanding tasks during peak energy periods and less critical or restorative tasks during low-energy periods. Each intentional choice reinforces efficiency, focus, and momentum. Momentum grows as students consistently work smarter, not just longer. Over time, proactive energy management produces discipline, strategic thinking, and high-impact achievement. Smart students recognize that thoughtful allocation of energy is more powerful than endless hours of effort.

This approach also enhances resilience in uncertain or high-pressure situations. Students who monitor energy can navigate challenges without becoming overwhelmed, maintaining effectiveness and emotional balance. Each decision to rest, recover, or shift effort reinforces adaptability and confidence. Momentum is maintained as students continue to perform even under demanding circumstances. Over time, this habit develops enduring resilience, focus, and sustained capability. Smart students convert energy awareness into a tool for handling complexity with grace and effectiveness.

Finally, managing energy instead of time supports long-term growth, achievement, and fulfillment. Students who optimize energy produce higher-quality work, sustain effort, and maintain motivation over years of learning and professional development. Each choice made to respect energy levels compounds into cumulative progress, skill, and confidence. Momentum grows as students experience the benefits of energy-aligned action consistently. Over time, this approach produces resilient, capable, and successful individuals. Smart students understand that energy, not hours alone, is the true currency of achievement, and managing it wisely is the key to a sustainable and fulfilling life.

QUOTES:

"Don't count the hours; invest your energy where it counts most."
— Anonymous

CHAPTER .38

BUILDING SELF-TRUST THROUGH FOLLOW-THROUGH

Many smart students struggle with self-trust because they fail to consistently follow through on commitments, promises, or goals. When actions don't match intentions, confidence erodes, and momentum stalls. Building self-trust through follow-through means creating habits of reliability with yourself first, completing tasks, and honoring promises no matter how small. Each commitment kept reinforces confidence, accountability, and internal momentum. Momentum grows as students develop a track record of dependability with themselves, which naturally extends to external responsibilities. Smart students understand that self-trust is built one completed action at a time.

One reason this approach is essential is that it strengthens confidence in decision-making. When students follow through consistently, they prove to themselves that they can act, persist, and achieve results. Each successful execution reinforces trust in judgment, ability, and discipline. Momentum is maintained as students approach new goals with assurance that they can follow through. Over time, this habit cultivates reliability, courage, and practical competence. Smart students convert consistent action into the foundation of self-trust and growth.

Building self-trust through follow-through also improves focus and prioritization. Students learn to identify commitments that matter most and allocate attention and effort accordingly. Each completed task reinforces the ability to focus on meaningful goals rather than being distracted by trivial or low-value activities. Momentum grows as students see consistent results from concentrated effort. Over time, this skill develops discipline, clarity, and purposeful action. Smart students leverage follow-through to align effort with impact, ensuring energy and focus are not wasted.

Another advantage is that follow-through strengthens resilience and perseverance. Students who commit to action and stick to it learn to overcome challenges, delays, and obstacles without giving up. Each instance of persistence reinforces self-trust, problem-solving, and momentum. Momentum is maintained as students develop confidence in their ability to navigate setbacks and continue forward. Over time, perseverance becomes habitual, creating a resilient mindset that supports consistent achievement. Smart students understand that honoring commitments builds both internal strength and external effectiveness.

Following through consistently also strengthens motivation and momentum. When students complete tasks, even small ones, they experience a sense of accomplishment that fuels further action. Each completed commitment reinforces confidence and energy to tackle the next challenge. Momentum grows as students accumulate evidence of their reliability and capability. Over time, consistent follow-through builds an internal drive that sustains effort without relying solely on external rewards. Smart students convert action into motivation, creating a self-reinforcing cycle of progress.

Another benefit is that follow-through builds clarity and reduces decision fatigue. Students who honor commitments avoid the mental burden of unfinished tasks and unresolved intentions. Each action completed clears space for focus on new priorities, reinforcing momentum. Momentum is maintained as students approach projects with a clear, organized mindset and renewed capacity. Over time, this habit cultivates strategic thinking, efficiency, and confidence in managing complex responsibilities. Smart students understand that self-trust grows when intentions are reliably converted into results.

Building self-trust through follow-through also improves resilience during setbacks. Students who have a track record of completing tasks are better able to navigate unexpected challenges because they trust their ability to persist. Each obstacle overcome while staying committed reinforces confidence and determination. Momentum grows as students tackle larger challenges with a proven sense of capability. Over time, resilience becomes habitual, supporting continuous learning and achievement. Smart students leverage follow-through as a foundation for enduring strength in the face of adversity.

This approach also enhances goal achievement and long-term success. Students who consistently act on commitments progress steadily toward meaningful objectives rather than relying on bursts of motivation or luck. Each completed step, no matter how small, compounds into measurable progress and sustained momentum. Momentum is maintained as students develop a reliable system of action that ensures results over time. Over time, this practice produces high-performing, accountable, and self-directed individuals. Smart students recognize that consistent follow-through is the bridge between intention and real-world accomplishment.

Following through consistently also strengthens self-discipline and focus. Students who honor commitments develop the habit of taking action even when they don't feel motivated or inspired. Each completed task reinforces the ability to work steadily toward goals despite distractions or fatigue. Momentum grows as students accumulate a record of reliable, focused effort. Over time, this habit cultivates discipline, attention, and a sense of control over personal progress. Smart students turn consistent action into a foundation for sustained achievement and growth.

Another advantage is that follow-through fosters confidence in tackling larger challenges. Students who build a track record of completing commitments gain assurance in their capacity to handle complex tasks and responsibilities. Each success, no matter the size, reinforces self-trust and momentum. Momentum is maintained as students approach new projects with experience and belief in their ability to deliver. Over time, this confidence becomes habitual, supporting ambitious goals and long-term growth. Smart students convert small acts of follow-through into evidence of competence and capability.

Building self-trust through follow-through also improves time and energy management. Students who honor commitments learn to prioritize tasks effectively and allocate energy where it matters most. Each fulfilled obligation provides clarity about what truly deserves focus, reinforcing momentum. Momentum grows as students maximize the impact of their effort through consistent action. Over time, this approach develops strategic thinking, efficiency, and sustained productivity. Smart students recognize that follow-through transforms intention into actionable, high-value outcomes.

This practice also strengthens resilience and adaptability. Students who follow through develop a sense of accountability that allows them to face setbacks without giving up. Each challenge overcome while maintaining commitment reinforces persistence, problem-solving, and momentum. Momentum is maintained as students continue forward, adjusting strategies and staying aligned with goals. Over time, this resilience becomes ingrained, supporting continuous progress and self-directed learning. Smart students leverage follow-through to turn obstacles into opportunities for growth and learning.

Following through consistently also reinforces integrity and authenticity. Students who honor commitments to themselves develop alignment between intentions and actions, creating a sense of personal honesty. Each action that matches intention strengthens self-respect, confidence, and momentum. Momentum grows as students accumulate evidence of their reliability and moral alignment. Over time, this habit fosters authenticity, ethical decision-making, and a stable sense of self. Smart students understand that self-trust is inseparable from the integrity of following through on their promises.

Another benefit is that follow-through encourages a growth mindset. Students who consistently act on commitments learn that effort, persistence, and consistency produce results over time. Each completed action reinforces the belief that progress comes from doing, experimenting, and learning, rather than waiting for perfection. Momentum is maintained as students approach new challenges with confidence that persistence leads to mastery. Over time, this habit cultivates resilience, adaptability, and continuous improvement. Smart students convert consistent follow-through into a framework for lifelong learning and achievement.

Building self-trust through follow-through also strengthens accountability and responsibility. Students who honor obligations to themselves develop a habit of ownership over their actions, decisions, and outcomes. Each fulfilled commitment reinforces personal responsibility and reduces reliance on external validation. Momentum grows as students internalize the link between effort and results, creating dependable patterns of action. Over time, accountability becomes habitual, producing self-directed, reliable, and capable individuals. Smart students recognize that responsibility begins with consistent action and honoring their own promises.

This approach also enhances clarity in goal achievement and personal development. Students who consistently follow through gain insight into their priorities, capabilities, and areas for growth. Each completed commitment provides feedback, reinforcing effective strategies and highlighting areas for improvement. Momentum is maintained as students refine approaches, maintain consistency, and advance toward meaningful goals. Over time, clarity becomes habitual, allowing students to act with purpose, focus, and efficiency. Smart students leverage follow-through as a tool for intentional, high-impact growth.

Following through consistently ultimately builds a strong foundation for lifelong self-reliance and success. Students who honor their commitments develop habits of reliability, accountability, and proactive action that compound over time. Each completed task reinforces trust in their own abilities, decision-making, and judgment. Momentum grows as students accumulate evidence of their competence, resilience, and consistency. Over time, this habit produces confident, capable, and self-directed individuals who can navigate challenges with assurance. Smart students recognize that the most important promises they can keep are the ones they make to themselves, and that consistent follow-through turns intention into achievement.

Another advantage is that follow-through strengthens emotional stability and reduces self-doubt. Students who reliably act on commitments feel empowered, capable, and in control of their progress. Each fulfilled obligation reassures them that they can handle challenges and meet goals, even when circumstances are difficult. Momentum is maintained as students approach new projects with confidence and composure. Over time, emotional stability becomes ingrained, supporting sustained focus, learning, and performance. Smart students turn consistent action into a tool for self-assurance and calm, strategic decision-making.

Building self-trust through follow-through also promotes resilience in the face of setbacks. Students who have developed the habit of honoring their commitments approach obstacles with persistence and problem-solving rather than avoidance or defeat. Each challenge navigated while staying committed reinforces adaptability, skill, and momentum. Momentum grows as students continue to advance despite difficulties, learning from experience rather than being derailed by it. Over time, resilience becomes habitual, producing capable, confident, and adaptable learners. Smart students convert follow-through into a strategy for turning challenges into opportunities for growth.

This approach also strengthens goal attainment and progress in meaningful work. Students who consistently act on priorities move steadily toward objectives, rather than waiting for perfect timing or external validation. Each completed step compounds into measurable outcomes, reinforcing momentum and motivation. Momentum is maintained as students build reliable patterns of action that produce results. Over time, this habit develops discipline, efficiency, and purposeful achievement. Smart students recognize that consistent follow-through bridges the gap between potential and tangible success.

Finally, building self-trust through follow-through cultivates a sense of personal empowerment and lifelong competence. Students who reliably honor their own commitments experience autonomy, confidence, and a track record of achievement. Each action taken reinforces the ability to make decisions, take initiative, and navigate uncertainty effectively. Momentum grows as students develop the self-trust needed to pursue ambitious goals with clarity and resilience. Over time, this practice produces self-directed, capable, and high-achieving individuals. Smart students understand that personal integrity and consistent follow-through are the most powerful tools for turning potential into enduring success.

Chapter .39

Turning Setbacks into Strategic Adjustments

Setbacks are inevitable for every student, no matter how smart or capable they are. The difference between stagnation and growth lies in how setbacks are approached. Turning setbacks into strategic adjustments means analyzing what went wrong, learning from the experience, and modifying actions to improve future results. Each setback becomes an opportunity to refine strategies, build resilience, and maintain momentum. Momentum grows as students shift focus from frustration to problem-solving and actionable lessons. Smart students understand that challenges are not failures—they are signals for adaptation and improvement.

One reason this approach is essential is that it transforms negative experiences into valuable feedback. Students who treat setbacks as learning opportunities extract actionable insights rather than becoming discouraged or giving up. Each lesson learned reinforces decision-making, planning, and skill development. Momentum is maintained as students continuously evolve strategies based on real-world outcomes. Over time, this mindset cultivates adaptability, analytical thinking, and resourcefulness. Smart students convert obstacles into stepping stones for progress and growth.

Turning setbacks into strategic adjustments also strengthens resilience and emotional control. Students learn to respond calmly and thoughtfully rather than reacting impulsively or emotionally. Each setback approached strategically reinforces confidence, patience, and persistence. Momentum grows as students practice maintaining focus and determination despite challenges. Over time, resilience becomes habitual, supporting sustained achievement and consistent performance. Smart students recognize that emotional composure is a critical skill in transforming adversity into opportunity.

Another advantage is that this approach improves problem-solving and critical thinking. Students who analyze setbacks can identify patterns, inefficiencies, and areas for improvement, creating actionable solutions. Each strategic adjustment enhances understanding, skill, and effectiveness. Momentum is maintained as students continuously refine methods and approaches for greater results. Over time, this habit develops foresight, adaptability, and innovative thinking. Smart students leverage setbacks as a source of insight, transforming obstacles into competitive advantages.

Analyzing setbacks also encourages strategic thinking and adaptability. Students who examine what went wrong can identify underlying causes and adjust their approaches accordingly. Each strategic adjustment reinforces problem-solving skills, judgment, and practical decision-making. Momentum grows as students learn to anticipate challenges and adapt methods before repeating mistakes. Over time, this practice develops flexibility, foresight, and a proactive mindset. Smart students turn setbacks into tools for strategic improvement rather than obstacles that halt progress.

Another advantage is that this approach reduces fear of failure. Students who view setbacks as opportunities to learn and adjust experience less anxiety and hesitation in pursuing ambitious goals. Each constructive response to a challenge reinforces confidence, courage, and persistence. Momentum is maintained as students approach future tasks with the mindset that mistakes are temporary, informative, and manageable. Over time, resilience and a growth-oriented perspective become habitual. Smart students convert setbacks into feedback loops that build confidence and accelerate achievement.

Turning setbacks into strategic adjustments also strengthens focus on what is controllable. Students learn to separate factors they can influence from external circumstances, directing effort where it matters most. Each adjustment grounded in controllable elements reinforces efficiency, clarity, and momentum. Momentum grows as students concentrate on actionable strategies rather than dwelling on unchangeable outcomes. Over time, this habit develops discipline, resourcefulness, and practical intelligence. Smart students understand that effective action comes from focusing on what they can change, not what they cannot.

This approach also fosters resilience in the face of repeated challenges. Students who consistently analyze setbacks and adjust strategies become less discouraged by repeated obstacles. Each problem navigated strengthens perseverance, adaptability, and strategic insight. Momentum is maintained as students continue moving forward despite adversity. Over time, this practice produces highly resilient, capable, and self-directed learners. Smart students leverage setbacks as a mechanism for growth, turning adversity into repeated opportunities for skill development.

Turning setbacks into strategic adjustments also strengthens problem-solving under pressure. Students who analyze challenges learn to develop solutions quickly and efficiently rather than becoming paralyzed by difficulty. Each successful adjustment reinforces confidence, creativity, and practical judgment. Momentum grows as students approach future obstacles with a mindset focused on solutions rather than problems. Over time, this habit fosters decisiveness, adaptability, and strategic thinking. Smart students transform pressure and setbacks into opportunities to improve processes and outcomes.

Another benefit is that this approach promotes a growth-oriented mindset. Students who consistently learn from setbacks recognize that intelligence, talent, and skill are developed through experience and reflection. Each challenge provides lessons that compound into greater knowledge and competence. Momentum is maintained as students internalize that progress results from persistent effort and adaptive thinking. Over time, this perspective produces resilience, continuous learning, and higher achievement. Smart students leverage setbacks as catalysts for development rather than viewing them as evidence of failure.

Turning setbacks into strategic adjustments also enhances emotional regulation and patience. Students who view obstacles as learning opportunities reduce frustration, stress, and impulsive reactions. Each setback approached thoughtfully reinforces composure, resilience, and strategic action. Momentum grows as students remain focused on long-term objectives rather than temporary setbacks. Over time, emotional stability becomes a natural response, supporting consistent progress and achievement. Smart students convert challenges into lessons that strengthen both mind and character.

This approach also improves long-term planning and foresight. Students who reflect on setbacks identify patterns and anticipate future obstacles, allowing for better preparation and strategic allocation of effort. Each insight gained from setbacks informs subsequent decisions, reinforcing momentum and effectiveness. Momentum is maintained as students take informed, proactive steps toward their goals. Over time, this habit develops foresight, efficiency, and higher-order thinking. Smart students turn setbacks into predictive tools that guide smarter action and sustained growth.

Turning setbacks into strategic adjustments also strengthens accountability and responsibility. Students who reflect on challenges learn to take ownership of their actions, decisions, and outcomes rather than blaming external factors. Each acknowledgment of mistakes or missteps reinforces self-trust, discipline, and momentum. Momentum grows as students implement changes that improve performance and prevent repeating errors. Over time, this habit fosters reliability, self-awareness, and proactive problem-solving. Smart students convert setbacks into actionable lessons that reinforce responsibility and growth.

Another benefit is that this approach encourages continuous learning and skill development. Students who strategically adjust after setbacks accumulate experience, insights, and strategies that compound over time. Each adjustment enhances knowledge, competence, and effectiveness. Momentum is maintained as students integrate lessons learned into new projects and challenges. Over time, this practice develops mastery, adaptability, and self-directed learning. Smart students leverage setbacks as opportunities to expand their skillset and refine approaches.

Turning setbacks into strategic adjustments also fosters resilience and confidence under uncertainty. Students who develop the habit of learning from setbacks gain assurance in their ability to navigate unpredictable situations. Each challenge faced and adjusted for reinforces courage, persistence, and momentum. Momentum grows as students tackle increasingly complex challenges with trust in their capability to adapt. Over time, resilience becomes a natural skill, supporting long-term achievement and sustained performance. Smart students transform obstacles into confidence-building experiences.

This approach also improves focus and prioritization. Students learn to distinguish between what is essential and what is irrelevant when responding to setbacks. Each strategic adjustment reinforces clarity, decision-making, and effective allocation of effort. Momentum is maintained as students concentrate on high-impact actions that produce meaningful results. Over time, this habit strengthens judgment, efficiency, and goal-oriented behavior. Smart students convert setbacks into opportunities to sharpen focus and direct energy where it matters most.

Turning setbacks into strategic adjustments ultimately transforms challenges into opportunities for mastery and growth. Students who consistently analyze setbacks, extract lessons, and implement changes develop a mindset focused on progress rather than perfection. Each adaptation reinforces momentum, skill, and confidence. Momentum grows as students convert obstacles into actionable insights that guide future decisions. Over time, this habit produces highly capable, resilient, and self-directed learners.

Smart students recognize that setbacks are not failures—they are essential tools for improvement, innovation, and lasting success.

Another advantage is that this approach strengthens perseverance and determination. Students who embrace setbacks as learning opportunities continue pursuing goals even when faced with difficulty or delay. Each challenge overcome while adjusting strategies reinforces persistence, discipline, and momentum. Momentum is maintained as students develop a reputation for reliability, resilience, and steady progress. Over time, perseverance becomes habitual, supporting consistent achievement and self-directed growth. Smart students leverage setbacks to cultivate determination that fuels long-term success.

Turning setbacks into strategic adjustments also improves self-awareness and reflection. Students learn to evaluate their strengths, weaknesses, and patterns of behavior when challenges arise. Each reflection generates insight, guiding better decisions and more effective strategies. Momentum grows as students continuously refine approaches based on experience and evidence. Over time, self-awareness becomes a core skill that enhances learning, adaptability, and judgment. Smart students transform setbacks into opportunities for self-knowledge and informed action.

This approach also fosters adaptability and innovation. Students who strategically adjust after setbacks become more flexible, creative, and willing to experiment with new methods. Each challenge approached with a problem-solving mindset reinforces the ability to pivot and innovate. Momentum is maintained as students explore alternative strategies that lead to better outcomes. Over time, this habit strengthens creativity, resourcefulness, and practical intelligence. Smart students convert setbacks into catalysts for experimentation, learning, and innovation.

Finally, turning setbacks into strategic adjustments creates a foundation for sustained growth and lifelong learning. Students who habitually analyze setbacks, implement improvements, and maintain momentum accumulate skills, confidence, and experience over time. Each strategic adjustment reinforces the ability to learn, adapt, and achieve consistently. Momentum grows as students develop a self-directed, proactive approach to challenges. Over time, this practice produces resilient, capable, and high-achieving individuals who thrive in any environment. Smart students understand that setbacks are opportunities for recalibration and growth, and turning them into strategic adjustments is the key to turning potential into lasting success.

QUOTES:

"Every setback is a setup for a stronger comeback." — Anonymous

CHAPTER .40

BECOMING A LIFELONG SELF-DIRECTED LEARNER

Becoming a lifelong self-directed learner is the ultimate skill smart students can develop to turn potential into lasting achievement. It means taking responsibility for your own growth, seeking knowledge and skills independently, and continuously adapting to new challenges. Each act of self-directed learning reinforces curiosity, discipline, and momentum. Momentum grows as students take initiative to pursue knowledge, apply it, and iterate on their understanding. Over time, this practice develops autonomy, resilience, and the ability to thrive in any environment. Smart students understand that learning how to learn is more powerful than any single skill or piece of information.

One key aspect of self-directed learning is goal setting and planning. Students identify what they want to learn, break it into actionable steps, and create a plan to achieve it. Each goal-oriented action reinforces clarity, focus, and momentum. Momentum is maintained as students consistently move toward meaningful objectives instead of waiting for external direction. Over time, this habit produces disciplined, purposeful, and strategic learners. Smart students convert intentional planning into a roadmap for continuous growth and achievement.

Becoming a lifelong self-directed learner also requires curiosity and a willingness to explore. Students who actively seek new ideas, perspectives, and experiences cultivate a mindset of continuous improvement. Each discovery and experiment reinforces adaptability, insight, and momentum. Momentum grows as students remain engaged, asking questions and testing their understanding. Over time, this mindset develops ingenuity, critical thinking, and a capacity to solve complex problems. Smart students leverage curiosity as fuel for ongoing personal and intellectual development.

Another critical component is reflection and self-assessment. Students who regularly evaluate their progress, strengths, and areas for improvement develop self-awareness and insight. Each reflective practice informs future actions and strategies, reinforcing momentum. Momentum is maintained as students adjust methods, correct mistakes, and refine skills continuously. Over time, reflection becomes habitual, strengthening judgment, adaptability, and self-directed learning. Smart students understand that growth accelerates when they actively assess their own performance and make informed adjustments.

Becoming a lifelong self-directed learner also strengthens problem-solving and critical thinking skills. Students who take initiative in their learning encounter diverse challenges and must develop solutions independently. Each challenge overcome reinforces reasoning, creativity, and momentum. Momentum grows as students build experience in analyzing situations and generating effective strategies. Over time, this habit cultivates intellectual flexibility, insight, and the ability to navigate complex problems. Smart students leverage self-directed learning to turn challenges into opportunities for skill and knowledge growth.

Another advantage is that self-directed learning fosters resilience and persistence. Students learn to continue seeking knowledge and improving skills even when progress is slow or obstacles arise. Each instance of persistence reinforces confidence, determination, and momentum. Momentum is maintained as students face difficulties with the mindset that effort and adaptation lead to mastery. Over time, this habit develops endurance, self-reliance, and long-term competence. Smart students convert setbacks into lessons, using resilience as a tool to sustain continuous growth.

Reflection and adaptation are also central to lifelong self-directed learning. Students must assess outcomes, identify gaps in knowledge, and adjust strategies to improve effectiveness. Each reflective practice reinforces learning, self-awareness, and momentum. Momentum grows as students continually refine their approaches and expand their understanding. Over time, this habit strengthens metacognition, strategic thinking, and adaptive intelligence. Smart students recognize that self-directed learning is iterative—progress comes from reflection, adjustment, and action.

Becoming a lifelong self-directed learner also enhances motivation and engagement. Students who take ownership of their learning pursue topics that matter to them, aligning effort with interest and curiosity. Each engagement with personally meaningful material reinforces enjoyment, focus, and momentum. Momentum is maintained as students find intrinsic motivation in pursuing challenges and learning new skills. Over time, this habit produces sustained engagement, creativity, and intellectual fulfillment. Smart students harness self-directed learning to cultivate passion alongside competence.

Becoming a lifelong self-directed learner also strengthens adaptability in unfamiliar situations. Students who independently pursue knowledge encounter diverse challenges that require flexible thinking and problem-solving. Each new situation navigated reinforces resourcefulness, creativity, and momentum. Momentum grows as students learn to adjust strategies, acquire new skills, and remain effective under changing circumstances. Over time, this habit develops versatility, confidence, and the ability to thrive in uncertainty. Smart students convert unfamiliar experiences into opportunities to expand competence and resilience.

Another benefit is that self-directed learning cultivates initiative and independence. Students who take charge of their own learning become proactive in seeking opportunities, solving problems, and acquiring skills without relying on external instruction. Each action taken independently reinforces self-trust, discipline, and momentum. Momentum is maintained as students gain confidence in their ability to direct learning and achieve goals autonomously. Over time, this habit produces self-reliant, motivated, and capable learners. Smart students understand that independence in learning leads to empowerment and long-term success.

Reflection and goal alignment are critical components of lifelong self-directed learning. Students who regularly assess progress, set objectives, and prioritize learning activities create a framework for effective, meaningful growth. Each alignment of action with purpose reinforces focus, clarity, and momentum. Momentum grows as students consistently make intentional decisions about what, when, and how to learn. Over time, this habit develops strategic thinking, intentionality, and sustained achievement. Smart students leverage reflection to ensure their learning is purposeful, impactful, and continuously advancing.

Becoming a lifelong self-directed learner also strengthens confidence in personal capabilities. Students who navigate challenges independently and achieve results build trust in their ability to learn, adapt, and succeed. Each successful application of knowledge or skill reinforces self-assurance and momentum. Momentum is maintained as students approach new learning opportunities with courage and initiative. Over time, confidence becomes habitual, empowering students to tackle complex problems and pursue ambitious goals. Smart students recognize that self-trust is reinforced through continuous, self-directed action and achievement.

Becoming a lifelong self-directed learner also strengthens problem ownership and accountability. Students who take responsibility for their own learning develop the habit of addressing challenges head-on and seeking solutions independently. Each instance of taking ownership reinforces self-trust, decision-making, and momentum. Momentum grows as students consistently apply their skills and knowledge to achieve meaningful results. Over time, this habit develops autonomy, reliability, and confidence in navigating complex situations. Smart students understand that owning their learning journey empowers them to shape outcomes and achieve their goals.

Another benefit is that self-directed learning promotes interdisciplinary thinking and integration of knowledge. Students who explore multiple fields, connect concepts, and apply insights across contexts develop broader perspectives and deeper understanding. Each connection made reinforces critical thinking, creativity, and momentum. Momentum is maintained as students leverage knowledge across areas to solve problems more effectively. Over time, this practice cultivates intellectual agility, innovation, and adaptable thinking. Smart students use self-directed learning to integrate diverse knowledge into actionable insight and skill.

Developing as a lifelong self-directed learner also strengthens initiative and proactive behavior. Students who seek challenges, identify opportunities, and act independently build the habit of moving forward rather than waiting for direction. Each proactive action reinforces confidence, capability, and momentum. Momentum grows as students create their own pathways to knowledge, skill, and achievement. Over time, initiative becomes habitual, supporting long-term growth and self-reliance. Smart students recognize that taking charge of their learning drives personal and professional success.

This approach also enhances resilience and persistence. Students who encounter obstacles while directing their own learning learn to navigate setbacks, adapt strategies, and persist until they achieve results. Each challenge overcome reinforces problem-solving ability, determination, and momentum. Momentum is maintained as students view setbacks as opportunities to learn and refine their approach. Over time, this habit produces resilience, endurance, and self-confidence. Smart students leverage self-directed learning to cultivate persistence and the ability to thrive under pressure.

Finally, lifelong self-directed learning fosters empowerment and mastery. Students who continually direct their own learning build skills, insight, and knowledge that compound over time. Each learning experience strengthens momentum, capability, and confidence. Momentum grows as students integrate lessons learned into increasingly complex tasks and goals. Over time, this practice produces highly skilled, adaptable, and confident individuals who can achieve their potential in any context. Smart students understand that the ability to learn independently is the most powerful tool for achieving mastery, growth, and fulfillment throughout life.

QUOTES:

"Education is not the filling of a pail, but the lighting of a fire." — W.B. Yeats

EPILOGUE

Smart students have a unique challenge: the world expects brilliance, speed, and flawless execution—but growth rarely follows a straight line. The most important lessons in life are often learned not from immediate success, but from struggle, reflection, and persistence. By now, you've explored mental traps, practical strategies, and ways to turn intelligence into momentum. You've learned how to embrace discomfort, build self-trust, and create a personal system for lifelong learning.

The next step is simple in theory but profound in practice: take what you've learned and act. Experiment. Fail. Adjust. Reflect. Repeat. Every step you take toward self-directed growth strengthens resilience, clarity, and mastery. This book cannot make decisions for you, but it can give you the frameworks, insights, and confidence to navigate your own path.

Remember: being smart is not a promise of ease—it's a foundation for ingenuity, courage, and growth. The power to move forward, to create a life aligned with your curiosity and values, is already in your hands. This is not the end of learning; it's the beginning of a lifelong journey of self-directed growth, momentum, and fulfillment.

About the Author

Laura Lee is a writer, creative entrepreneur, and digital visionary whose work bridges innovation and human connection. Raised in a quiet town where stories were lifelines, she discovered early the transformative power of expression and reflection.

As the founder of a successful digital design agency, Laura has helped countless brands craft identities that reflect clarity, purpose, and authenticity. Her writing—both personal and universally resonant—echoes the same principles that guide her work: intention, growth, and conscious creation.

Laura believes that intelligence is a tool, not an identity, and that meaningful progress comes from curiosity, persistence, and self-trust. Through her books, platforms, and consulting, she empowers readers and creators to embrace their potential, navigate challenges, and design lives aligned with their values and strengths.

This book is an extension of that mission: a guide to self-discovery, a roadmap for moving from stuck to momentum, and a companion for anyone seeking clarity, confidence, and the courage to own their learning journey.

www.ingramcontent.com/pod-product-compliance
Lightning Source LLC
LaVergne TN
LVHW052304200926
848481LV00002B/458